SpringerBriefs in Applied Sciences and Technology

For further volumes:
http://www.springer.com/series/8884

Elisa Di Giuseppe

Nearly Zero Energy Buildings and Proliferation of Microorganisms

A Current Issue for Highly Insulated and Airtight Building Envelopes

 Springer

Elisa Di Giuseppe
Dipartimento DICEA
Università Politecnica delle Marche
Ancona
Italy

ISSN 2191-530X   ISSN 2191-5318   (electronic)
ISBN 978-3-319-02355-7   ISBN 978-3-319-02356-4   (eBook)
DOI 10.1007/978-3-319-02356-4
Springer Cham Heidelberg New York Dordrecht London

Library of Congress Control Number: 2013949182

Printed on acid-free paper

Springer is part of Springer Science+Business Media (www.springer.com)

Preface

The proliferation of microorganisms in buildings has an enormous aesthetic, health and economic impact. However, to date few studies have highlighted the possible worsening of this problem in relation to the construction methods of "Nearly Zero Energy Buildings".

This brief book wants to give a small original contribution to putting emphasis on this topic, in an attempt to raise public awareness towards a way of constructing buildings which is not just energy efficient (toward "Nearly Zero Energy") but which is also sustainable, safe and provides high living quality for building occupants.

Research on these topics is very broad and multidisciplinary: it involves aspects of microbiology for the classification of the main infesting organisms, building physics issues related to the environmental conditions for the proliferation, aspects of materials engineering with regard to the biodeterioration of materials and remedial actions, medicine for what concerns the consequences for human health. The preparation of the book has therefore been made possible thanks to the collaboration of experts in these different fields.

This will be just a starting point for further investigation, so that building construction of the future not only focuses on energy efficiency but also on the quality of life of the inhabitants.

Acknowledgments

The multidisciplinary topic of this book has necessitated the cooperation of experts in different fields.

Chapter 6 has been written in collaboration with Prof. Marco D'Orazio, whose expertise in building physics was substantial for the definition of thermo-physical phenomena that regulate the heat and moisture transport in NZEB and their relationship with biological growth.

Prof. Enrico Quagliarini gave a significant contribution in Chap. 7, for the description of remedial actions and future innovations.

The author is very grateful to them.

The author wishes also to acknowledge Prof. Francesca Clementi, Drs. Lucia Aquilanti and Andrea Osimani for their assistance, support and scientific guidance in writing Chaps. 3 and 4.

Contents

Chapter 1
Definitions

1.1 Nearly Zero Energy Building (NZEB)

There are several definitions of what a "Zero Energy Building" means in practice, with differences in usage all around the world. In this book, we will refer to the definition of the European Directive on the Energy Performance of Buildings (EPBD recast 2010/31/EU). "Nearly Zero Energy Building" means "*a building that has a very high energy performance* [...]" and in which "*the nearly zero or very low amount of energy required should be covered to a very significant extent by energy from renewable sources, including energy from renewable sources produced on-site or nearby*". Since EPBD recast does not provide harmonised requirements and details of energy performance calculation framework, it will be up to the European Member States to define what "*a very high energy performance*" and "*a very significant extent by energy from renewable sources*" mean.

1.2 Building Envelope

The building envelope is the physical barrier that separates the interior of the building from the outdoor environment, with the principal functions of structural support and control of heat, air and moisture flows. It usually consists of all of the exterior components of the building, including walls, roofing, foundations, windows and doors.

1.3 Thermal Insulation

The thermal insulation of the building envelope is the means for achieving a reduction of heat transfer between the interior of the building and the outdoor environment due to a difference in temperature. It is usually obtained by adding adequate layers of materials with very low thermal conductivity (W/mK) to the

E. Di Giuseppe, *Nearly Zero Energy Buildings and Proliferation of Microorganisms*, SpringerBriefs in Applied Sciences and Technology, DOI: 10.1007/978-3-319-02356-4_1, © The Author(s) 2013

building envelope, thereby obtaining a building envelope with low U-value (thermal transmittance, W/m^2K) or high R-value (thermal resistance, m^2K/W).

"Superinsulation" or "Overinsulation" is an approach to building design, construction and retrofitting that radically reduces heat loss (and gain) by using much higher levels of insulation and airtightness than normal (U-values ranging from 0.15 to 0.10 W/m^2K).

1.4 Airtightness

The airtightness of a building is the procedure of "sealing" the envelope to reduce air leaks through possible gaps in walls, roofs, windows, etc… and to consequently reduce the heat transfer by convection. This often requires significant quantities of insulation and very airtight windows.

The less natural airflow into a building, the more mechanical ventilation will be required to support human comfort and indoor air quality (IAQ).

1.5 Thermal Decoupling

"Thermal decoupling" is the complete breaking between the thermal behaviour of the inner part of the envelope (completely influenced by the interior conditions) and the external part (subject to climatic conditions) due to the high insulation of building envelope. This phenomenon can cause an overheating of the external layers of the envelope during the warmer seasons and an "undercooling" effect during the colder ones, above all in night time.

1.6 ETICS

ETICS mean External Thermal Insulation Composite Systems. In general, an ETIC consists of an insulation board fixed to a building wall with adhesives, anchors or mechanically with rails. It is coated with reinforced plaster and then finished by a finishing plaster or other coating materials.

1.7 Sick Building Syndrome

The term "sick building syndrome" (SBS) is used to describe situations in which building occupants experience acute health and comfort effects that appear to be linked to time spent in a building, but no specific illness or cause can be identified.

Building occupants complain of symptoms associated with acute discomfort, e.g. headache; eye, nose or throat irritation; dry cough; dry or itchy skin; dizziness and nausea; difficulty in concentrating and fatigue. Most of the complainants report relief soon after leaving the building. The main contributing factors cited for the SBS are: inadequate ventilation, chemical contaminants from indoor or outdoor sources, biological contaminants.

1.8 Bioreceptivity

Bioreceptivity refers to the aptitude of a material to be biologically colonised and is related to material properties that contribute to the anchorage and development of different types of organisms.

1.9 Biofilm

A "biofilm" is a microbial coating on material surfaces in which cells stick to each other and are frequently embedded within a self-produced matrix of extra-cellular polymeric substance. Biofilms can contain many different types of microorganisms, e.g. bacteria, archaea, protozoa, fungi and algae; each group performs specialised metabolic functions. The formation of a biofilm begins with a weak reversible adhesion of the microorganisms to a surface. Then, if the colonists are not immediately separated from the surface, they can anchor themselves more permanently using cell adhesion structures.

1.10 Biodeterioration

"Biodeterioration" is any undesirable change of the structure and the stability of a material due to the activity of living organisms. It is usually divided into:

- Physical or mechanical biodeterioration (disruption of the material).
- Biochemical biodeterioration, due to the fact that the organisms use the material as source of nutrients and energy or that they excrete metabolic products that can damage the material.
- Aesthetic biodeterioration, simply due to the presence of a biofilm.

It is common to find all three types of biodeterioration acting together.

Chapter 2
Introduction

The thermal performance of the building envelope is central to the debate on the construction of "Nearly Zero Energy Buildings" (NZEB). The relationship between type of envelope and final energy consumptions is usually translated into a simplistic deduction: to ensure that the energy consumption of a building is close to zero, it is primarily necessary to dissipate very low heat during the cold season. A global building thermal resistance and airtightness is one of the most important prerequisites to achieve "nearly zero energy".

Consequently, several countries have increased their airtightness and thermal resistance requirements in buildings, and the construction market is always more oriented towards "overinsulated" lightweight envelopes and a global reduction of air permeability of windows. Also renovation techniques aim to obtain the same prerequisites, for instance by replacing single glazed windows by new very tight double or triple glazed windows, or by adding interior or exterior insulation.

However, poorly permeable buildings are more subject to high internal moisture load in combination with an unsuitable ventilation strategy. Modern exterior insulation finish systems do not have thermal inertia and are therefore more subject to considerable amounts of exterior condensation.

Moisture loads and surface condensations are favourable conditions for the proliferation of microorganisms, such as algae and fungi. Consequently, despite the fact that over the past decades building energy efficiency has improved and better quality in living spaces is required, the number of reports on the presence of microorganisms on building facades and indoors is still increasing.

The proliferation of microorganisms in buildings is not welcome not only because of the implications for human health but also because of their contribution to the defacement of materials. The danger for the building occupants lies in the spread of pathogens (disease causing agents). In comparison to health aspects, other building damages caused by microorganisms—i.e., the aesthetic defacement and biodeterioration, is of minor importance. However, even these latter issues should not be excluded from a more thorough investigation. In fact, always more often, the biodeterioration of building materials leads to social and economic disruption for residents and builders.

E. Di Giuseppe, *Nearly Zero Energy Buildings and Proliferation of Microorganisms*,
SpringerBriefs in Applied Sciences and Technology, DOI: 10.1007/978-3-319-02356-4_2,
© The Author(s) 2013

The NZEB of the future must be able to give a concrete answer to these problems, but the research on the relationship between biological proliferation and NZEB is still in its infancy. It certainly needs further investigations in the light of the real actual data collected.

The problem is vast and complex and requires a multidisciplinary approach. In fact, it involves aspects of building physics, material engineering, microbiology and medicine. A deep insight is necessary in the short term, because the housing market progresses quickly, adapting to a constantly changing regulatory framework.

In the first part of this book (Chaps. 3 and 4), some information is provided on the principal proliferating microorganisms in contemporary buildings (algae and mould), their causes and conditions of growth, and the resulting consequences for the building materials and people's health.

In Chap. 5, some of the major biological risk prediction models have been reported. Many building hygrothermal analysis methods are able to simulate the coupled transport processes of heat and moisture for one or multidimensional cases which aim to predict biological risks. Additional measurements in laboratory and in situ conditions are often used for the validation of the models. At the end of the chapter, some methods of accelerated experimental testing for the evaluation of biological defacement of buildings materials have then been described.

Since the recent construction practices that aim to reach the NZEB standard are likely to create more favourable conditions for biological growth—as mentioned in Chap. 6, by a description of real cases in literature—controlling and preventing solutions become more and more pressing and important.

Chapter 7 provides an overview of both traditional and new strategies. Traditional strategies include mechanical, physical and above all chemical methods, useful to eliminate the presence of microorganisms and, when possible, to delay their recurrence. Other researches are looking towards safer and more sustainable strategies linked to a proper design, choice of materials and construction of buildings, aiming to obtain a good envelope performance without any added costs.

Chapter 3
Algal Growth on External Building Envelope

Abstract Algae are very ancient living organisms. Their presence on earth came about some 3.5 billion years ago. They are considered "pioneer organisms" of outdoor environment, and it is actually possible to find different varieties of algae on the ground, in the air, in ice and even in anthropogenic elements such as the facades of buildings since they are able to survive through frequent freeze–thaw and dehydration cycles. The aesthetic quality and durability of an external building envelope could be seriously impaired by the development of algae which will colonise building materials whenever a suitable combination of humidity, warmth and light occurs. The fundamental role of water for algal growth is clear which, for several reasons, is found in large quantities on building facades. External sources of water here include rain, snow, ground moisture, airborne humidity and condensation of vapour from outdoor air. In addition to environmental conditions, the rate of stain development largely depends on the "bioreceptivity" of the material, that is, its aptitude to be biologically colonised which is related to the material properties that contribute to the anchorage and development of microorganisms. The facades of the buildings are then fertile substrates for the growth of algae.

Keywords Algae · Cyanobacteria · Biofilm · Condensation · Driving rain · Biodeterioration · Bioreceptivity

3.1 Most Common Varieties of Algae Proliferating on Building Envelope

Most of the algae known live in freshwater (rivers and lakes) and only 10 % of them live in the sea. It is possible to find different varieties of algae on the ground, in the air, in ice and even in anthropogenic elements such as the facades of buildings.

In particular, green algae and cyanobacteria (mistakenly called "blue algae", "blue-green algae" or "Cyanophyta") are a phylum of photosynthetic bacteria.

E. Di Giuseppe, *Nearly Zero Energy Buildings and Proliferation of Microorganisms*,
SpringerBriefs in Applied Sciences and Technology, DOI: 10.1007/978-3-319-02356-4_3,
© The Author(s) 2013

They are unicellular or pluricellular organisms which may be eukaryotic (green algae) or prokaryotic (cyanobacteria). One important characteristic that distinguishes algae from mould fungi is their autotrophy, which is the ability of an organism to synthesise its own organic molecules starting from inorganic substances, using energy that has not been derived from assimilated organic substances. Therefore, the algae, by means of chlorophyllian photosynthesis (Eq. 3.1), transform light energy into chemical energy and they are also able to synthesise inorganic compounds such as carbon dioxide, water and some elements, and thereby obtain organic substances that guarantee their long-term survival:

$$CO_2 + H_2O + \text{light energy} \rightarrow CH_2O + O_2 + H_2O \\ + \text{heat (Photosynthesis)} \qquad (3.1)$$

The community of organisms on building envelopes usually forms a "biofilm". A biofilm is an extremely complex microbial ecosystem that may consist of diverse amounts of different microorganisms together with a matrix of organic and inorganic nutrients and extracellular polysaccharide substances secreted by the cells.

Formation of a biofilm often begins with the attachment of autotrophic organisms to a surface. These first colonists adhere to the surface initially through weak, reversible forces, and if they are not directly separated from the surface, again they can anchor themselves more permanently using cell adhesion molecules. The first colonists facilitate the arrival of other organisms by building a matrix that holds the biofilm together. It is the biofilm matrix rather than the organisms that is in immediate contact with the surface of the building material (Johansson 2006).

Kappock (2004) divided the colonisation of microorganisms in buildings into two stages. "Primary infestation" (up to 6 months after painting) is a complex interaction of paint formulation, substrate, climate and local microorganisms and depends on rainfall, film porosity, type and quantity of leachable organic material (thickeners, dispersants, wetting agents). "Secondary infestation" occurs when paint material breaks down and/or nutrient materials are deposited from the air, changing the exposed surface.

Biofilm can be found on any building. It can be located in a polluted area or not. It has traces of different colours (black, green, red, etc...) and may extend over the entire façade (Fig. 3.1). It usually takes at least one year to appear on walls, but once the favourable growth conditions are met, the development may be dazzling (Dubosc 2000).

Classifying the algae is very complex because first of all the criteria of differentiation are sometimes very difficult to identify, and secondly, according to their vegetative state (growth or survival) they may assume very different aspects.

Fig. 3.1   Example of biofilm with traces of different colours (*black, green, yellow*) extend over the entire surface of a building façade

For these reasons, in the literature there are different classifications. We will make reference to the classification of the cryptogams indicated by Dusbosc (2000):

1. Prokaryotic organisms (no nucleus)

   (a) no chlorophyll → bacteria
   (b) presence of chlorophyll → cyanobacteria (blue-green algae)

2. Eukaryotic organisms (with nucleus)

   (a) no chlorophyll → fungi
   (b) presence of chlorophyll → cyanobacteria (blue-green algae)

– rootless → algae
– symbiotic algae—fungus → lichens
– with roots → foams.

A cryptogam is a plant (in the wide sense of the word) that reproduces by spores, without flowers or seeds. The names of all cryptogams are regulated by the International Code of Nomenclature for algae, fungi and plants (ICN) (McNeill et al. 2012). The two categories of Cryptogams commonly found on the facades of the buildings are cyanobacteria and microalgae.

3.1.1 Cyanobacteria

They are improperly called "blue-green algae" and "Cyanophyta" and are a phylum of bacteria that obtain their energy through photosynthesis. Their cells are surrounded by a well-defined membrane that is often coloured. They have a homogeneous content that is slightly lighter towards the centre, with occasional rare and minor punctuation. The set of cells is within a translucent substance, inside the coloured part, which can be very thin (Fig. 3.2).

The colonisation of building materials by these organisms is encouraged by some of their intrinsic characteristics:

- vital reactivity at low light intensity;
- their membranes and their mucilage may be impregnated with colours that protect them against strong sunstroke;
- some of them can fix nitrogen gas into ammonia (NH_3), nitrites (NO_2) or nitrates (NO_3) which can be absorbed and converted to protein and nucleic acids;
- hygroscopicity of the cell envelope allows to retain water and dissolved salts;
- the presence of an outer envelope and the mucilage favour the biological state of latency in situations of prolonged drought.

3.1.2 Microalgae

Microphytes or microalgae are unicellular species, which exist individually, or in chains or groups. Depending on the species, their sizes can range from a few micrometres (μm) to a few hundreds of micrometres.

Their cells, although more complex than those of blue-green algae, contain different organelles, including one or more chloroplasts, recognisable by their clearly defined outlines and their green colour. Devoid of accessory pigments and

Fig. 3.2 Scheme of the cell of a cyanobacterium

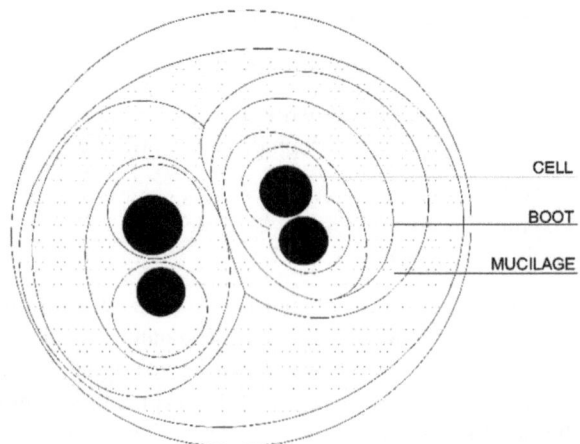

colouring or protective slime, they can only live in environments where humidity is constant and sufficient illumination is present without it being too intense. In these favourable conditions, they proliferate extremely fast, and they can turn a white wall green in a few days. They are also often in a vegetative state in the middle of blue-green algae that protect them.

Diagnosis made on cyanobacteria and microalgae, through an optical microscope (magnification X400) of scraped samples (Grossin and Dupuy 1978) allowed the identification of the main types present on the building envelope (Tables 3.1 and 3.2).

The biological growth on limestone walls sees the predominant presence of *Cyanophyceae* (blue-green algae), secondarily associated with other plants. Identified species vary following the exposure of the wall to the sun, but many species especially from genera *Gloeocapsa* and *Chroococcus* can be found (Dubosc 2000).

Studies of stone monuments from different continents with black dirt show the presence of abundant microscopic algae. Species that we encountered fall into two

Table 3.1 Main kinds of cyanobacteria recorded on building envelopes (Dubosc 2000)

Cyanobacteria
The cells have a rounded contour and a homogeneous content except for a slight variation of colour and for the presence of small grains.

Aphanocapsa
Round, yellow, green or light brown cells, grouped in a translucent gel that can be easily seen since they are well defined
Notes: They develop at low light intensity

Aphanothece
Elongated cells with colour from greenish yellow to light brown rounded at both ends and enclosed in a transparent jelly
Notes: They develop at low light intensity

Gloeocapsa
Rounded and sometimes hemispherical cells with brownish or reddish haloformation by mucilage
Notes: They prefer sunny places

Gloeothece
Cells with rounded shape, green or brown, with formation of halos by mucilage
Notes: They are spread all over the world

Chroococcus
Cells which are rounded, oval and sometimes irregularly shaped. They have a lot of brown slime
Notes: They are found only in very sunny places

Nostoc
Round, light coloured, green or brown cells, with the presence of mucilage
Notes: They require adequate lighting

Calothrix
Brown or greenish cells, grouped into filaments with one tapered end and one rounded end
Notes: They prefer sunny places

Scytonema
The cells are grouped into filaments, are brownish or greenish, and have a branched form
Notes: They prefer sunny places

Table 3.2 Main kinds of microalgae recorded on building envelopes (Dubosc 2000)

Microalgae
The cells have a net wall and a content of cells in which there are one or more well-defined large particles (chloroplasts).
Chlorella, Chlorhormidium, etc.
Green algae, *Chlorococcacées, Ulothricacéees*
Notes: They are often found in places with high humidity, along with cyanobacteria, mosses and lichens
Trentepohlia
Orange cells with thick walls and granular contents grouped into filaments
Notes: They prefer sunny places

broad classes. First, cyanobacteria whose genres represented are: *Phormidium, Gloeocapsa, Chroococcus* and secondly, *Chlorophyceae: Chlorhormidium, Trentepohlia, Chlorella*. They are present mainly due to high humidity conditions and their water retention surface (Zelia Almeida De França and Miller 2010).

A study of the brick structures (mineral medium) in Belgium (Guillitte 1998) shows that, in addition to mosses and lichens, the dominant fouling organisms observed are algae, especially cyanobacteria (*Lyngbya, Gloeocapsa, Anabaena*) and green algae (*C. chlorhormidium*).

Sometimes, the dominant species change depending on the material, as noted by Schlichting (1975), who observed *Chlorella* (green alga) develop on bricks, while the mortar joint was colonised by *Chroococcus* and *Schizotrix* (cyanobacteria). As for the materials described above, the presence of moisture in large enough quantities mainly triggered their development.

On painted walls, various species have been observed, including cyanobacteria *Lyngbya* and *Gloeocapsa*, and green algae *Chlorella* and *Trentepohlia* (Gaylarde et al. 2011; Gaylarde and Gaylarde 2000).

Biofilm presence on concrete has been detected in many areas of the world. Developments of microscopic algae have been observed on concrete in England, due to the presence of high humidity because of surface runoff or water stagnation. Reddish developments disfigure concrete walls throughout Ireland and are mainly composed of a green alga *Trentepohlia*. In Singapore, *chlorophyceae Trentepohlia* was identified on concrete with red dirt and black streaks. In France, on several concrete walls, with or without paint, most spills are caused by microscopic algae. The colours are varied and the colour encountered depends directly on the type of algae observed. Black streaks mainly consist of blue-green algae, while green and red stains are caused by *Chlorophyceae* (Dubosc 2000).

3.2 Main Causes and Conditions of Growth

Algae are "pioneer organisms" of outdoor environment and are widespread on the external surfaces of buildings since they are able to survive through frequent freeze–thaw and dehydration cycles. In fact, they are particularly resistant to wind and rainfall, although they are not well-protected against evaporation. For this reason, they require more moisture than moulds (present in the surrounding environments).

It is well established that the growth of algal species is favoured by an optimal combination of biotic and abiotic factors. Especially among them there are: nutrients coming from the substrate; type of organisms present; moisture and other environmental factors (Fig. 3.3). If any one of these three macro-conditions is missing, vegetative growth becomes unlikely.

For photosynthesis, sufficient light, water, temperature, carbon dioxide and some mineral nutrients must be present. In general, green algae develop with RH = 70–80 %, while blue algae with RH = 100 % and a temperature between 15 and 50 °C, even if the range 20–25 °C is the ideal temperature (Zillig et al. 2003).

Some algae for their growth also need some trace elements (as Fe, Mn, Si, Zn, Cu, Co, Mo, B, V), which are available in our environment because of either run-off rainwater or pollution.

From the definition of the conditions for development and growth of microalgae and cyanobacteria, the fundamental role of water is clear which, for several reasons is found in large quantities on the facades of buildings. External sources of water include rain, snow, ground moisture and airborne humidity.

The main moisture loads on façade systems are wind-driven rain and condensation of vapour from outdoor air (Künzel 2010). The facades of the buildings are then fertile substrates for the growth of algae, as we will see in the following sections.

Fig. 3.3 Schematic representation of the optimal conditions for the growth of algae on building materials

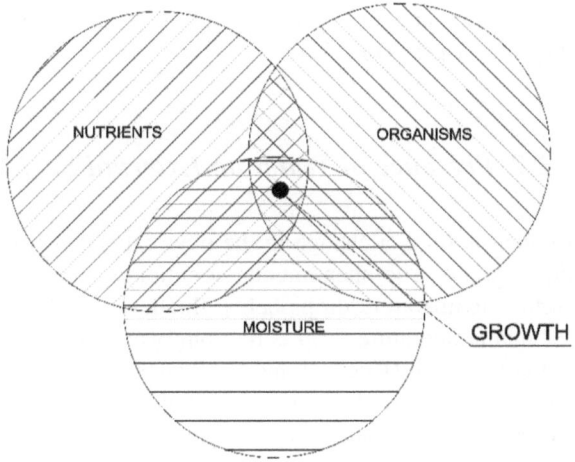

3.2.1 Presence of Water on the External Surfaces:
The Driving Rain

The driving rain load on buildings has been studied extensively (Blocken 2004). The presence of water is a natural condition on the external surfaces of buildings. Its propagation depends on both geometric factors (slope and configuration of the surfaces) and on the physical characteristics of the support.

Moreover, whether an exposed façade is prone to microbial growth also depends on the duration as well as the intensity of the driving-rain spells and on the drying conditions between these spells.

It has been demonstrated that porosity, roughness and composition of the external finishes all participate in retaining rain (Barberousse et al. 2007): roughness influences adhesion; porosity determines absorption by capillarity; mineral composition is crucial to provide nourishment to algae.

Even the shape of the building may create preferential routes where the rain-water stagnates, creating ideal conditions for the growth of microorganisms. While a roof overhang can reduce the driving-rain load on the façade, a slight tilt of the wall may increase it. This is a rather serious problem for heritage buildings with skew envelope parts or modern architecture with deliberately inclined façades. Once the algae have grown, the run-off rainwater contributes to replacing the old cells with new cells and favours the spread of spots of biofilm to other non-contaminated building components (Künzel 2010).

According to Knight and Hammett (1993), there are two ways in which water comes in contact with wet masonry constructions. First, rain, usually wind-driven, falls directly on vertical, horizontal and inclined surfaces. Second, rainwater runs over and off these surfaces. In the latter case, the direction of water run-off depends on the inclination of the surface, on the texture of the material, absorption, degree of saturation and on surface configurations. In this case, water patterns often have a fingering aspect (Fig. 3.4).

3.2.2 Presence of Water on External Surfaces: Condensation

Condensation is a change in the physical state of matter from gaseous phase into liquid phase. The phenomenon commonly occurs on building facades when their surface temperature drops below the dew point[1] of ambient air. The main reason for this temperature drop is the long-wave radiation exchange of the façade with the atmosphere (Kuenzel and Sedlbauer 2001), which results in a net heat flux to the sky during night-time depending on the emissivity and temperature of the surfaces. The phenomenon has a maximum effect during clear and cold nights

---

[1] Dew point is the temperature below which the water vapour in a volume of humid air at a given constant barometric pressure will condense into liquid water.

Fig. 3.4 A typical algal growth on a balcony caused by wind-driven rain. The water run-off carries dirt and spores from the balcony slab causing the formation of typical vertical stripes

when condensing conditions can persist on the surfaces for up to 15 hours. Künzel (2010) showed that the condensation formed during the night on the surfaces of vertical walls exposed to critical fronts determines the exceed for a significant time the condition of relative humidity of 80 %. This limit is defined as *TOW80* (Time of Wetness) and represents a period of time, in terms of hours per day, when surface relative humidity remains above 80 % (Adan 1994).

3.2.3 Additional Environmental and Climatic Factors

Climate, the surrounding environment, exposure, season and ventilation conditions will affect the level of algal growth occurring in a region. The interactions are complex and interwoven. Industrial emissions and tree cover, for example, affect both solar penetration and temperature at the same time. Location and climate influence the microbiota and chemical content of the air hitting exposed surfaces (Gaylarde et al. 2011).

In the Northern hemisphere, deterioration due to algal species is found mainly on north and northwest facing surfaces (Nay and Raschle 2003), since these are hardly ever irradiated by the sun throughout the day and remain damp for a longer time (Fig. 3.5). Walls that have a good exposure to the sun receive more light and more heat and therefore dry out in a short time. Consequently, they are less prone to microbial growth. Direct sunlight may also damage the photosynthetic pigments

Fig. 3.5 Traces of *grey* biofilm over the northern façade of a building in Italy exposed to wind-driven rain

of the algae: the optimal light intensity for algae is about 1,000 lux, usually corresponding to north-facing walls (Karsten et al. 2005).

Barberousse et al. (2007) found that variables related to humidity are the most important for determining the presence of most phototrophic species and suggested that north-facing façades are more highly colonised by phototrophs because they are cooler and dry more slowly after rain.

Even the season is important for the formation of algae on the façade. In fact, during the hot season the external surfaces tend to dry out more quickly, hindering the development of these organisms. However, if algae develop when conditions are more favourable, the season is decisive for their survival. In summer the external surface temperature of buildings may reach temperatures close to 60 °C, causing the death of the algal cells (Karsten et al. 2005), while in winter, temperatures lower than 15 °C inhibit their development.

Künzel et al. (2006) argue that the seasonal periods that facilitate the growth and development of algae are the intermediate seasons, fall and spring. The spring season is privileged compared to the summer season because it has high peaks of temperature and rainfall has greater loads. Instead, the fall season is more conducive than winter because it has less rigid temperatures, higher rainy loads and greater light. In particular, during the autumn season, temperatures permit more likely the formation of condensation on the surfaces.

Macroclimate, the meteorological conditions at the site, may be less important for microbial colonisation than microclimate. Shirakawa et al. (2010) found that the environment and climatic conditions were the main factors affecting colonisation, rather than biocide presence and paint formulations. Paint films exposed to the tropical climate of São Paulo (Brazil), influenced by the South Atlantic Convergence Zone and facing south, developed biofilms containing the greatest number of viable fungi. In the Southern hemisphere, south facades receive less solar radiation than those facing the north, and their surface temperatures may fall below dew point at night and remain moist for longer periods after wetting. The equatorial climate of Belém (Brazil), which has high solar radiation and rainfall, favoured colonisation by cyanobacteria. The temperate climate of Rio Grande, in the coastal zone of southern Brazil associated with high irradiation, did not facilitate significant photroph colonisation, being the only region studied in which the white paint maintained its initial colour after a 4 year exposure.

3.2.4 Influence of Technical and Manufacturing Solutions

In addition to environmental conditions, the rate of microorganism development in buildings largely depends on the "bioreceptivity" of the material. Bioreceptivity, which refers to the aptitude of a material to be colonised by living organisms, is related to material properties and to the technical and construction choices of the buildings themselves, which can interact directly or indirectly (Table 3.3).

The geometrical characteristics of the building facades are very important for the ability of water (rainwater, condensation, capillary rise) to stagnate on the surface and enhance the development and the growth of algae. A high thermal insulation of the building envelope can accentuate the phenomenon (see 6.1.1).

Furthermore, great importance is coated by the finishing materials of the facades, whose physic-chemical characteristics may determine substrates more or less favourable to the emergence and proliferation of microorganisms. The latter are in fact deposited on surfaces from the surrounding environment. Once they impinge on the surface, they adhere and grow at rates that depend on the nature of the coating, the substrate and the environmental conditions. The constituents of the

Table 3.3 Main construction choice that could affect the microorganism development on a building surface

Construction choice	Influence on microorganism development
Architectural form	Presence of water
Thermal insulation	Condensation
Finishing material	Suitable substrate
Geographical location	Climatic factors
Presence of vegetation	Climatic factors

Fig. 3.6 Microorganism colonisation of a clay brick façade. Clay bricks usually have great roughness and porosity, so they are highly bioreceptive

coating affect microbial development, some components being inhibitory and others stimulatory to growth (Gaylarde et al. 2011).

Phototrophic organisms, the cyanobacteria and algae, can flourish also without recourse to organic nutrients and are found in the majority of external paint biofilms.

Mortar is a mixture of sand and a powdered adhesive, such as cement and water. Once applied as paste, it dries up and becomes hard, forming pores of different sizes that creates suitable microhabitats for many types of microorganisms. Mortars thus possess high primary bioreceptivity and microorganisms may cause the powdering of material (Urzì and De Leo 2007).

Physical parameters of coatings such as roughness or porosity are strongly supposed to be factors of influence (Fig. 3.6). Accelerated laboratory tests of biological growth on mortar samples that were performed (Dubosc et al. 2001; Kuenzel and Sedlbauer 2001) show that algal developments increase with the porosity of the underlying material. Thus, it seems that the use of dense, high-performance mortars can slow down or even inhibit microorganism growth.

It is also known that finish materials with high water absorption are more prone to biological colonisation (Kuenzel and Sedlbauer 2001).

Barberousse et al.(2007), evaluated the susceptibility of various types of external façades to algal growth by an accelerated water-streaming test method (see 5.2). The experiments were conducted with algae isolated from biofilms developing on real building facades in France. The tested materials were a laboratory-made mortar and four manufactured products (two one-coat rendering mortars, one organic finish and one paint), selected because they were representative of the facade coatings marketed in Europe.

One-coat rendering mortars are mineral materials applied in one coat which fulfils all the functions of a multi-coat rendering system used externally and which is usually specifically coloured. The tested mortars were ready-mixed dry materials composed of sand, cement, lime, mineral pigments and specific admixtures.

Organic finishes and paints are very common products for outdoor applications and mainly used for decorative purposes. In particular, organic finishes are known to be widely used as final coats of ETICS.[2] The organic finish and the paint of the tests were water-based and ready-to-use acrylic-based products. The organic finish also contained mineral aggregates.

The test results showed that the organic coatings are more resistant to colonisation than mortars. Among the mineral coatings, the colonisation kinetics also seemed to be linked to roughness and porosity, as mortar, which was the most rapidly colonised and the most porous and roughest. Moreover, the organic finish was more colonised than the paint, which may be due to its higher roughness.

Venzmer et al. (2008) carried out long-term natural weathering tests with the aim of identifying factors of algae growth on external surfaces. To this aim, they made numerous samples of ETICS with different surface finishes and subjected them to natural environmental conditions for a period of 4 years (from October 2003 to 2007). The main surface finishes analysed were synthetic resin plaster, silicon plaster, silicate plaster, standard mineral plaster, pebble dash plaster, thick-film plaster.

The tests performed (without any initial biological presence) have shown that it takes at least 18 months prior to obtaining algae visible to the naked eye. Furthermore, construction systems more susceptible to the algal development are those that employ synthetic-based resin or silicone plasters.

All paints contain same primary constituents: a matrix or binder, pigments and extenders (which confer colour and build) and a solvent. The solvent is either organic in nature, for a solvent-based paint, or water, for a water-based or latex emulsion paint. In addition to these common components, there are several other ingredients, which make up about 5 % of the total system. Paints can also be classified as either being alkyd or acrylic in nature. An alkyd paint is usually oil-based and is made from a synthetic resin that is made by reacting a drying oil with a hard synthetic material. An acrylic paint is usually water-based. The binder in this case is made from a synthetic polymer (Gaylarde et al. 2011).

---

[2] External Thermal Insulation Composite Systems.

Cellulosic components can act as nutrients for microorganisms and inorganic materials. Leaching from the paint, such as phosphates, can aid algal growth on the surface (Kappock 2004).

Dark-pigmented filamentous cyanobacteria (mainly *Fischerella/Mastigocladus* or *Scytonema*) have been shown to be the principal component of thin black films on painted surfaces in Campeche, Mexico and Belize City (Gaylarde and Gaylarde 2000).

Pigment volume content (PVC), the volumetric ratio between pigment and resin, given as a percentage, is an important paint parameter, determining gloss and permeability. The relationship between PVC and susceptibility of paint films to biofilm growth is complex (Gaylarde et al. 2011). Reports of the effect of PVC on biosusceptibility vary, sometimes stating that this parameter is of relatively little importance (Shirakawa et al. 2010), sometimes indicating it as fundamental to paint film biodeterioration (Wagner 2001).

Organic pigments are more likely to act as microbial nutrients. Additionally, impurities in the pigments, such as phosphates and potassium salts, may act as essential micronutrients, increasing paint susceptibility to biodeterioration.

Recently, Breitbach et al. (2011) studied the rate of fouling of painted fibre cement panels exposed for 34 months in Florianopolis, southern Brazil. Out of the ten differently pigmented acrylic paints, blue, red and ceramic colours were, in that order, the slowest to become discoloured. Resistance to fouling may have been due to copper in blue and acidity from sulphur oxides in ceramic pigments.

3.3 Consequences for Durability and Performance of Building Elements

Paints and other surface coatings confer two main properties to a building envelope: protection and decoration. These properties may be impaired by microbial growth (Gaylarde et al. 2011).

"Biodeterioration" is defined as any undesirable change in the properties of a material caused by the vital activity of any organism (Morton and Surman 1994).

Biodeterioration was estimated as long ago as 1981 to cause economic losses in excess of one million US dollars per year (Winters 1981), and modern buildings, with increased levels of thermal insulation, show a higher risk of microbial growth (Sedlbauer et al. 2011) (see 6.1).

The metabolic activities of microorganisms, such as the production of extra-cellular polymers, the liberation of chelating compounds and of organic/inorganic acids, together with the presence of coloured pigments and the mechanical pressure exerted by growing structures or shrinking/swelling phenomena, induces different types of damages:

Fig. 3.7 Evident damages on
a building surface due to
microorganism colonisation

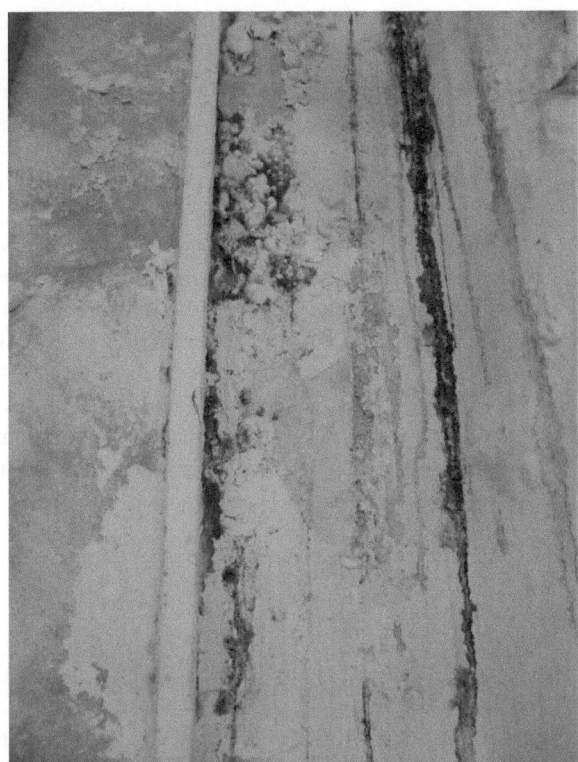

- physical (abrasion, mechanical stress);
- chemical (solubilisation of essential minerals, excretion of organic acids or enzymes, pH reduction, change of the electrical conductivity);
- aesthetical (coloured patches or patinas and crusts).

The consequences on building surfaces are the formation of biofilms, coloured patinas, encrustations and the presence of vegetative and reproductive bodies. Moreover, the coating can be subject to dwindling, erosion, pitting, ion transfer and leaching processes (Fig. 3.7).

The intensity of these damages are strictly correlated with: type and dimension of the organism involved; kind of material and state of its conservation; environmental conditions, micro-climatic exposure; level and types of air pollutants (Tiano 2002).

Cyanobacterial and algae on stone surfaces are considered "biodeteriogenic" because of the aesthetic damage they cause, producing a variety of coloured patinas (Ariño et al. 1995). Two other types of biodeterioration can also be considered: biogeophysical and biogeochemical.

Biogeophysical deterioration can be defined as the mechanical damage caused by exerted pressure during biological growth and through cycles of drying and moistening, resulting in surface detachment, superficial losses or penetration and increased porosity. Furthermore, the formation of biofilm results in longer moisture retention on the surface, increasing the potential of colonisation.

Biogeochemical deterioration is the direct action caused by the metabolic processes of organisms on the substratum because of the release of corrosive acids which contribute to salt formation. Furthermore, the microbial biofilms favour the adherence of airborne particles (dust, pollen, spores, carbonaceous particles from combustion of oil and coal), giving rise to crusts and patinas.

There are many other references in the literature that point to direct decay mechanisms caused by photosynthetic microorganisms (Crispim et al. 2006; Peraza Zurita et al. 2005; Warscheid and Braams 2000).

However, some researchers have doubted to what extent the organisms actually damage the façades on buildings (Prieto Lamas et al. 1995; Sterflinger 2000). Some studies even show that growth of lichens can actually protect the façade (Ariño et al. 1995; Bjelland and Thorseth 2002) or compensate for environmental stresses (Edwards et al. 1997; Edwards et al. 1995).

References

Adan OCG (1994) On the fungal defacement of interior finishes. Dissertation, Eindhoven University of Technology

Ariño X, Ortega-Calvo J, Gomez-Bolea A, Saiz-Jimenez C (1995) Lichen colonization of the Roman pavement at Baelo Claudia (Cadiz, Spain): biodeterioration vs. bioprotection. Sci Total Environ 167:353–363

Barberousse H, Ruot B, Yéprémian C, Boulon G (2007) An assessment of façade coatings against colonisation by aerial algae and cyanobacteria. Build Environ 42:2555–2561. doi:10.1016/j.buildenv.2006.07.031

Bjelland T, Thorseth IH (2002) Comparative studies of the lichen–rock interface of four lichens in Vingen, western Norway. Chem Geol 192:81–98. doi:10.1016/S0009-2541(02)00193-6

Blocken B (2004) Wind-driven rain on buildings. Dissertation, Castholic University of Leuven, Belgium

Breitbach AM, Rocha JC, Gaylarde CC (2011) Influence of pigment on biodeterioration of acrylic paint films in Southern Brazil. J Coat Technol Res 8:619–628. doi:10.1007/s11998-011-9350-1

Crispim CA, Gaylarde PM, Gaylarde CC, Neilan BA (2006) Deteriogenic cyanobacteria on historic buildings in Brazil detected by culture and molecular techniques. Int Biodeterior Biodegradation 57:239–243. doi: 10.1016/j.ibiod.2006.03.001

Dubosc A, Escadeillas G, Blanc PJ (2001) Characterization of biological stains on external concrete walls and influence of concrete as underlying material. Cem Concr Res 31:1613–1617. doi: 10.1016/S0008-8846(01)00613-5

Dubosc A (2000) Etude du développement de salissures biologiques sur les parements en béton: mise au point d'essais accélérés de vieillissement. Dissertation, Institut National des Sciences Appliquées de Toulouse

Edwards H, Russell N, Seaward M (1997) Calcium oxalate in lichen biodeterioration studied using FT-Raman spectroscopy. Spectrochim Acta Part A Mol Biomol Spectrosc 53:99–105. doi:10.1016/S1386-1425(97)83013-2

Edwards H, Russell N, Seaward M, Slarke D (1995) Lichen biodeterioration under different microclimates: an FT Raman spectroscopic study. Spectrochim Acta Part A Mol Biomol Spectrosc 51:2091–2100. doi:10.1016/0584-8539(95)01499-1

Gaylarde CC, Morton LHG, Loh K, Shirakawa MA (2011) Biodeterioration of external architectural paint films—a review. Int Biodeterior Biodegradation 65:1189–1198. doi: 10.1016/j.ibiod.2011.09.005

Gaylarde PM, Gaylarde CC (2000) Algae and cyanobacteria on painted buildings in Latin America. Int Biodeterior Biodegradation 46:93–97. doi:10.1016/S0964-8305(00)00074-3

Grossin F, Dupuy P (1978) Méthode simplifiée de détermination des constituants des salissures: Proceedings du Colloque International RILEM sur l'Altération et la Protection des Monuments en Pierre, Paris, France, p 41

Guillitte O (1998) Bioreceptivity and biodeterioration of brick structures. In: Conservation of historic brick structures: case studies and reports of research, Donhead Publishing, Shaftesbury, pp 68–84

Johansson S (2006) Biological growth on mineral façades. Dissertation, Lund University, Sweden

Kappock P (2004) Biocides: wet state and dry film. In: Handbook of coating additives, Marcel Dekker, New York

Karsten U, Schumann R, Haubner N, Friedl T (2005) Lebensraum fassade: aeroterrestrische mikroalgen. Biol unserer Zeit 35:20–30. doi:10.1002/biuz.200410269

Knight T, Hammet M (1993) The interaction of design and weathering on masonry constructions. Masonry International 7:9–13

Kuenzel H, Sedlbauer K (2001) Biological growth on stucco. In: Proceedings of the 8th international conference on performance of exterior envelopes of whole buildings, Clearwater Beach, Florida, pp 1–5

Künzel HM (2010) Factors determining surface moisture on external walls. In: Proceedings of the 11th international conference on thermal performance of the exterior envelopes of whole buildings, Clearwater Beach, Florida

Künzel HM, Künzel H, Sedlbauer K (2006) Hygrothermische beanspruchung und lebensdauer von wärmedämm-verbundsystemen. Bauphysik 28:153–163. doi:10.1002/bapi.200610015

McNeill J, Barrie F, Buck W (2012) International Code of Nomenclature for algae, fungi, and plants (Melbourne Code). In: Proceedings of the 18th international botanical congress Melbourne, Australia

Morton L, Surman S (1994) Biofilms in biodeterioration—a review. Int Biodeterior Biodegradation 34:203–221. doi:10.1016/0964-8305(94)90083-3

Nay M, Raschle P (2003) Wie lassen sich Algen und Pilze an Fassaden verhindern. Tagungsband, 131–138

Peraza Zurita Y, Cultrone G, Sánchez Castillo P, et al (2005) Microalgae associated with deteriorated stonework of the fountain of Bibatauín in Granada, Spain. Int Biodeterior Biodegradation 55:55–61. doi: 10.1016/j.ibiod.2004.05.006

Prieto Lamas B, Rivas Brea MT, Silva Hermo BM (1995) Colonization by lichens of granite churches in Galicia (northwest Spain). Sci Total Environ 167:343–351. doi:10.1016/0048-9697(95)04594-Q

Schlichting HE (1975) Some subaerial algae from Ireland. Brit Phycol J 10:257–261. doi:10.1080/00071617500650251

Sedlbauer K, Krus M, Fitz C, Künzel H (2011) Reducing the Risk of Microbial Growth on Insulated Walls by PCM Enhanced Renders and IR Reflecting Paints. In: Proceedings of the 11th DBMC international conference on durability of building materials and components, Porto, Portugal, pp 1–7

Shirakawa MA, Tavares RG, Gaylarde CC et al (2010) Climate as the most important factor determining anti-fungal biocide performance in paint films. Sci Total Environ 408:5878–5886. doi:10.1016/j.scitotenv.2010.07.084

Sterflinger K (2000) Fungi as geologic agents. J Geomicrobiol 17:97–124. doi:10.1080/01490450050023791

Tiano P (2002) Biodegradation of cultural heritage: decay mechanisms and control methods. In: Proceedings of the 9th ARIADNE workshop on historic material and their diagnostic, ARCCHIP, Prague

Urzì C, De Leo F (2007) Evaluation of the efficiency of water-repellent and biocide compounds against microbial colonization of mortars. Int Biodeterior Biodegradation 60:25–34. doi:10.1016/j.ibiod.2006.11.003

Venzmer H, Von Werder J, Lesnych N, Koss L (2008) Algal defacement on facade materials—results of a long term natural weathering tests obtained by new diagnostic tools. In: DTU (ed) Proceedings of 8th symposium on building physics in the nordic countries, Copenhagen, Denmark, pp 277–284

Wagner O (2001) Sauber bleiben: Anschmutzungsverhalten von wässrigen Fassadenfarben. Farbe + Lack 107:105–134

Warscheid T, Braams J (2000) Biodeterioration of stone: a review. Int Biodeterior Biodegradation 46:343–368. doi:10.1016/S0964-8305(00)00109-8

Winters H (1981) Latex Paints. Academic Press, Oxford UK

Zelia Almeida De França A, Miller AZ (2010) Primary bioreceptivity of limestones from the mediterranean basin to phototrophic microorganisms. Dissertation, Universidade Nova de Lisboa

Zillig W, Lenz K, Sedlbauer K, Krus M (2003) Condensation on the facade. Influence of construction type and orientation.In: Proceedings of the 2nd international conference on building physics, Antwerpen, Belgium, pp 437–444

Chapter 4
Development of Mould in Indoor Environments

Abstract Mould spores are widely disseminated in the environment. Even inside the buildings, we can find hundreds of species of fungi, proliferating with a favourable combination of conditions (oxygen, appropriate temperature, moisture, nourishment from the substrate) in which to germinate, grow, and sporulate. The presence of mould in buildings is not welcome for two main reasons: they are responsible for several types of illnesses and pathologies experienced by building occupants, grouped under the name of "Sick Building Syndrome", and their presence contributes to the defacement of paint and finishes. In recent years, since buildings are always more airtight and highly insulated, internal moisture load risks to become always greater if not managed by an adequate strategy. Consequently, the presence of moulds has considerably increased, despite the fact that living spaces should have better quality. Attention should be paid to the correct application of all that leads to minimisation of mould risk in buildings.

Keywords Mould · Spore · Germination · Condensation · Water activity · Substrate · Bioreceptivity · Sick building syndrome

4.1 Mould Life Cycle

Mould represents all species of microscopic fungi that grow in the form of multicellular filaments, called *hyphae*. In contrast, microscopic fungi that grow as single cells are called *yeasts*. A connected network of tubular branching hyphae has multiple, genetically identical nuclei and is considered a single organism, referred to as a colony.

Fungi are ubiquitous eukaryotic organisms, comprising an abundance of species. The main characteristic of mould fungi is that they do not have chloroplasts and therefore they are not able to carry out photosynthesis.

Over 1.5 million of fungal species exist on earth, 65,000 of which have been identified. Although they are all slightly different, mould species that grow in

E. Di Giuseppe, *Nearly Zero Energy Buildings and Proliferation of Microorganisms*,
SpringerBriefs in Applied Sciences and Technology, DOI: 10.1007/978-3-319-02356-4_4,
© The Author(s) 2013

Fig. 4.1 Representation of
the life cycle of mould

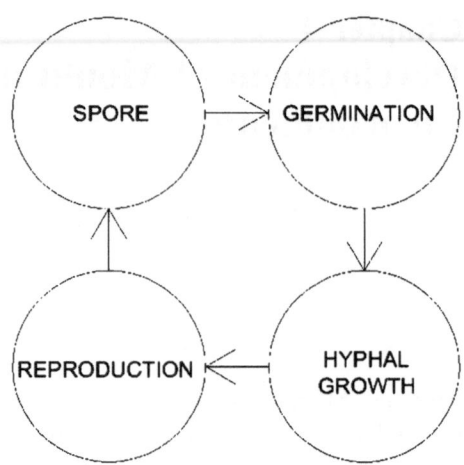

building spaces exhibit a similar life cycle. The typical life cycle of fungi consists of four stages: spore, germination, hyphal growth (vegetative growth), and reproduction (Moon 2005).

During the germination phase, the spores settle on surfaces and remain inactive until they can absorb moisture and nutrients from the substrate. If the substrate does not provide adequate nourishment and moisture, the spores do not germinate. The growth of hyphae occurs immediately after germination and as these thicken, they form a mass called *mycelium*. From this time onwards, the fungi metabolise the substrate material by extracting the necessary nutrients and retaining the moisture needed for growth. In the final phase, the fungi constitute a reproductive organism producing spores (Fig. 4.1).

Concerning mould growth in buildings, it is often categorised into visible or non-visible states (Moon 2005). Viitanen et al. have developed a seven-level mould growth index (Viitanen et al. 2000). Sedlbauer et al. used another index for mould growth intensity with six levels (Sedlbauer et al. 2003) (Table 4.1).

From the study on the life cycle of moulds, it emerges that the mould needs adequate nourishment from the substrate and moisture for its growth. The control of environmental conditions in buildings is then critical for mould prevention, as reported in the following sections.

4.2 Main Causes and Conditions of Growth

The problem of mould growth inside buildings has been observed in different geographical areas and various types of building (Daquisto et al. 2004) (Fig. 4.2).

Fungi may be transported into buildings on the surface of new materials or on clothing. They may also penetrate buildings through active or passive ventilation.

Table 4.1 Indices of fungal growth: the seven-level mould growth index on the *left* (Viitanen et al. 2000); the six-level index on the *right* (Sedlbauer et al. 2003)

Index of mould development	Growth rate	Intensity of growth	Characteristics
0	No growth (inactive spores)	0	No detectable growth
1	Growth observed only with a microscope (early stages of development of the hyphae)	1	Growth visible only with a microscope
2	Moderate growth detected under the microscope (covering more than 10 %)	2	Growth visible to the naked eye
3	Moderate growth detected visually (new spores produced)	3	Remarkable growth
4	Growth detected visually (covering more than 10 %)	4	Strong growth
5	Abundant growth detected visually (covering more than 50 %)	5	Total growth
6	Dense growth (covering over 100 %)	–	

Fungi are therefore found in the dust and surfaces of every house, including those with no problems with damp.

To proliferate, mould requires a favourable combination of environmental conditions in which to germinate, grow, and sporulate (Moon 2005). In the next section, we will go deeper into all these conditions.

4.2.1 Environmental Factors

Environmental conditions for fungal spore settling in buildings, include oxygen, appropriate temperature, and moisture or water activity (Hens 1999). Water activity (aw) is a measure of water availability and is defined as the ratio of the vapour pressure of the material pore (p) to that of pure water (p_o) at the same temperature ($Aw = p/p_o$). It takes values between 0 (dry substance) and 1 (pure water). In the substrate, the presence of many different components makes the relationship between humidity and water activity extremely complex and nonlinear. It is described by the curve of the water adsorption isotherm. The fungal growth on building materials is therefore linked to the curve of water adsorption of the material constituting the substrate.

Minimum water activity required for fungal growth on building surfaces varies from less than 0.80 to greater than 0.98 (Grant et al. 1989). Based on their water requirements, indoor fungi can be divided into:

(1) primary colonisers, which can grow at a water activity less than or equal to 0.80;

(2) secondary colonisers, which grow at a water activity level of 0.80–0.90;

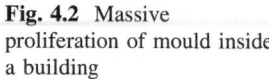

Fig. 4.2 Massive proliferation of mould inside a building

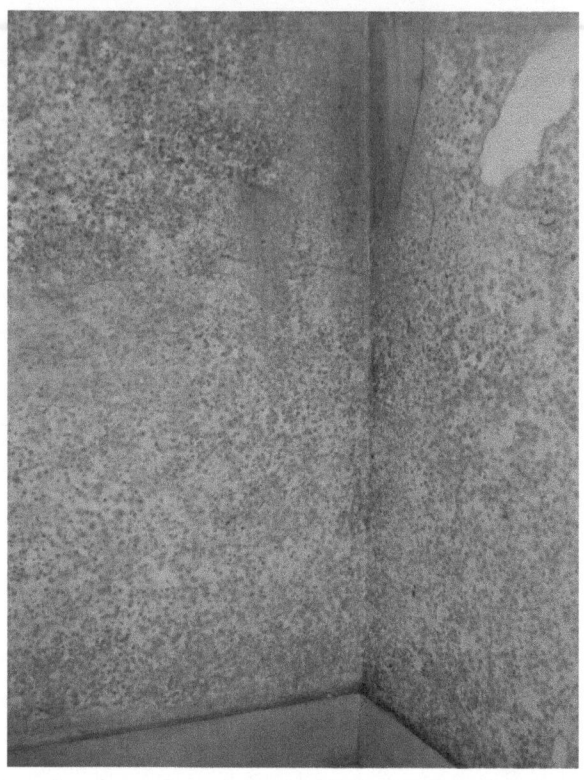

(3) tertiary colonisers, which require a water activity greater than 0.90 to germinate and start mycelial growth (Grant et al. 1989).

High levels of humidity, some surface and interstitial condensation may be sufficient for most primary and secondary colonisers, while tertiary colonisers generally require higher condensation levels, in combination with poor ventilation, or water damage from leaks, flooding and groundwater intrusion (Haseltine and Rosen 2009).

As most indoor fungi grow at 10–35 °C, common indoor temperatures are also not a limiting factor. However, although temperature and nutrients are not critical, they may affect the rate of growth and the production of certain allergens and metabolites (Nielsen et al. 2004).

For their life indoors, Fungi also need nutrients, which may include carbohydrates, proteins and lipids, coming from indoor environments. Moreover, they can even grow on inert materials such as ceramic tiles and can obtain sufficient nutrients from dust particles and soluble components of water.

From the studies in this field, a clear uniformity in defining the following four distinctive basic factors for the formation of mould is seen: temperature, moisture, nutrients (due to the type of support) and exposure time (Fig. 4.3).

Mould fungi especially grow on materials in the following specific conditions:

- presence of oxygen;
- temperatures between 22 and 35 °C (Baughman and Arens 1996);
- indoor relative humidity ranging between 71 and 95 % (Ayerst 1969);
- adequate substrate to provide the nutrients (Hens 1999).

Although each fungal species has a preferential humidity of growth, which varies according to the temperature and time of persistence of favourable environmental conditions, the International Energy Agency indicates an average of 80 % RH as a critical threshold for the formation of moulds (IEA 1990).

Other secondary factors for mould growth are: the pH value and the roughness of the substrate on which moulds grow, light, biotic interactions between different cultures, and indoor air velocity (Adan 1994; Krus et al. 2001).

4.2.2 Influence of the Type of Support

Most fungi are saprophytes, which means that they can feed on carbohydrates, proteins and lipids. In indoor environments, sources are varied and abundant: plants, pets, dust and building materials (such as wallpaper and fabrics),

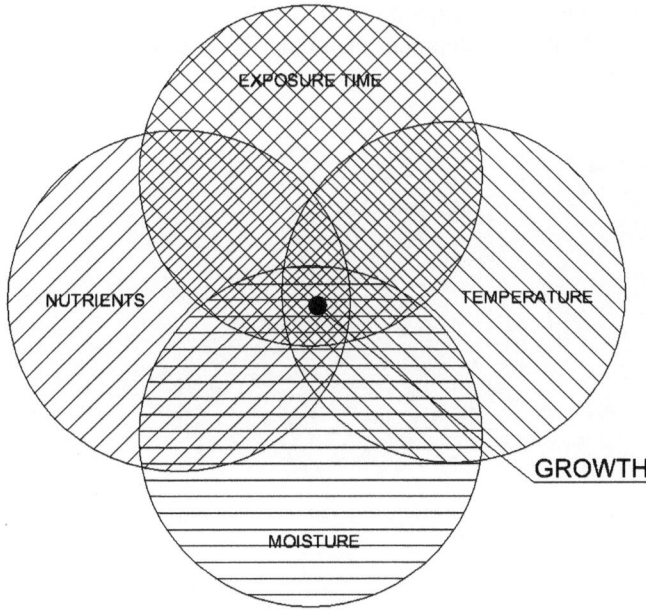

Fig. 4.3 Requirements for fungal growth indicated by various authors

condensation/deposit of cooking vapours, paint and glue, wood, packed products (such as food), books and paper objects.

Several authors have studied the relationship between construction materials and mould growth. However, the results available are not easily resumable because of the great variety of materials and other factors that intervene to influence the development of moulds (Fig. 4.4).

Fungi are able to grow on almost all natural and synthetic materials, especially if they are hygroscopic or wet. Inorganic materials get frequently colonised as they absorb dust and serve as good growth substrates for *Aspergillus fumigatus* and *Aspergillus versicolor* (Haleem Khan and Mohan Karuppayil 2012). Wood is highly vulnerable to fungal attack. *Cladosporium* and *Penicillium* (*Penicillium brevicompactum* and *Penicillium expansum*) are reported to infest wooden building materials. Dried wood surfaces are more susceptible to fungi (Sailer et al. 2010). Acylated wooden furniture, wood polyethylene composites, plywood and modified wood products are susceptible to infestation by *Aspergillus*, *Trichoderma* and *Penicillium*. Inner wall materials used in buildings, such as prefabricated gypsum board, highly favour the growth of *Stachybotrys chartarum*. Gypsum supports fungal growth, as it is hygroscopic. Paper and glue used in indoor surfaces are very good growth substrates for most of the indoor fungi (Haleem Khan and Mohan Karuppayil 2012).

Aspergillus and *Penicillium* grow superficially on painted surfaces, and *Aureobasidium pullulans* have been found to deteriorate the paints (Shirakawa et al. 2002). Acrylic painted surfaces are attacked by *Alternaria*, *Cladosporium* and *Aspergillus* (Shirakawa et al. 2011).

Sedlbauer provides the following clear classification of substrates in relation to their sensitivity to the development of mould (Krus et al. 2001; Sedlbauer et al. 2001):

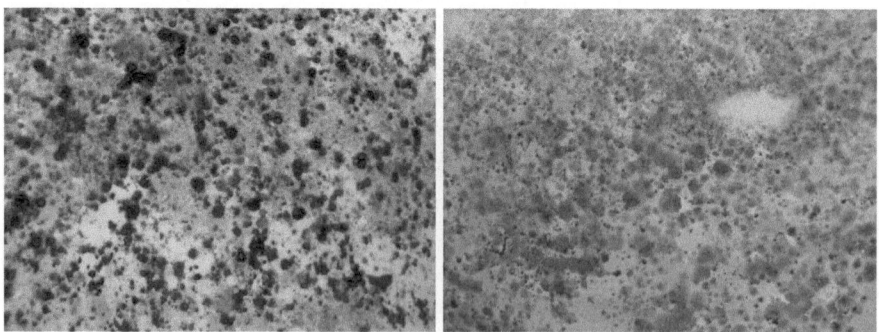

Fig. 4.4 Development of different types of moulds on the same substrate (internal mortar of a building)

Substrate category 0: Optimal culture medium;

Substrate category I: Biologically recyclable building materials like wall paper, plaster cardboard and building materials made of biologically degradable raw materials, material for permanently elastic joints;

Substrate category II: Building materials with porous structure such as renderings, mineral building material, certain types of wood as well as insulation material not covered by I;

Substrate category III: Building materials that are neither degradable nor contain any nutrients.

On the basis of these categories, the author developed an "isopleth system" (see Sect. 5.1) which allows the determination of the timing of germination of spores and growth of mycelium on the basis of limit curves of temperature and relative humidity (Krus et al. 2001).

Some building materials which are particularly rich in carbon, like cellulose or carbonates (wallpaper, wood-based building materials), are more conducive to the development of moulds compared with others that have lower carbon content (for example, plaster and glass wool) (Górny et al. 2003; Vacher et al. 2010).

Some paint components encourage microbial growth on the surface. Cellulosic components can act as nutrients for microorganisms that produce cellulase enzymes.

Vacher et al. studied the susceptibility to mould growth for various coatings and paint (acrylic-based painting, paint containing glycerine, wallpaper ordinary 80 g/m^2 and a wallpaper thick 240 g/m^2), applied on three types of substrates (Vacher et al. 2010):

- non-biodegradable;
- biodegradable and those treated with biocidal products;
- biodegradable and those free from biocides.

The outcome was that on aluminium and biocide-containing or untreated plasterboard surfaces, both acrylic-based and glycerol-based paints led to fungal growth, while unpainted aluminium and biocide-containing plasterboard were relatively or completely resistant. Researchers argued that the composition of the coatings and paints is as important as the substrate in making predictions about the biological degradation caused by mould. Therefore, the resistance of the vertical walls to mould contamination must refer to the exact composition of the surface coating (paint, wallpaper, etc..).

Ritschkoff et al. conducted experiments on wood-based composite materials (chipboard planks, wood wool and plywood boards), plaster, concrete, insulating materials (glass wool and rock wool) in different temperature and relative humidity conditions (Ritschkoff et al. 2000). Their results show that all the materials in a building may contribute to the growth of moulds if their relative humidity reaches 90 %. Some inorganic materials, such as metals and plastic, are not in themselves nutrients that are suitable for mould fungi, although the dust that deposits on them may represent a source of nourishment.

D'Orazio et al. showed that there is a direct correlation between the growth rate of some fungal species (*S.chartarum, P.Chrisogenum and A. versicolor*) and the content in organic matter, which various coatings and indoor finishes are able to provide as nutrients for the spores (D'Orazio et al. 2009). The authors underlined that, although various types of plaster and finish (experimentally analysed) belong to the same class of substrate (II), according to Sedlbauer's classification (Sedlbauer 2001), in reality there are sometimes quite remarkable differences in the results for the various substrates. Experimental results showed that the species *S. chartarum* (the most toxic for human health) had the most widespread development on the various types of support surface used.

Viitanen et al. developed a mathematical model for determining risk and durability of various materials to fungal growth under different conditions (Viitanen et al. 2011). They underlined that perhaps the most difficult requirement of the calculation is assessing the sensitivity of the material. Künzel states that "The ultimate goal of building physics related durability research should be the development of material specific degradation models" (Künzel 2011).

4.3 Effects on Human Health

Many people live more than 80 % of their time indoors; consequently, the indoor air quality (IAQ) has a fundamental effect on comfort, health and productivity, and has become a health priority for children according to the European Environmental Agency (Tamburlini et al. 2002).

The concentration of microorganisms inside buildings is often higher than that in the atmosphere and is responsible for several types of illnesses and pathologies experienced by the occupants and link with the time spent indoor, grouped under the name of "sick building syndrome" (SBS). Headaches, pressure on the head and throbbing and feelings of tiredness are the most common signs of SBS (Haleem Khan and Mohan Karuppayil 2012).

Numerous epidemiological studies have demonstrated that long-term exposure in unhealthy environments, subject to the proliferation of moulds and fungi, is one of the main causes of allergies and irritative reactions. Engvall et al. have shown that SBS is most frequent in places with a high level of humidity together with pungent odours and moulds (Engvall et al. 2002). Araki et al. has studied the possible causal relationship between SBS and indoor air quality (Araki et al. 2012). They describe how the pathology is found more often in environments characterised by the presence of moulds. This phenomenon is mainly due to the facility with which spores and their metabolic waste can be inhaled or ingested.

In many types of fungi, the spores are only 2 μm in size and can therefore easily penetrate the bronchial tubes. Spores that have a diameter of greater than 10 μm cannot arrive at the bronchial cavities but are retained in the mucous membranes of the pharynx and may give rise to allergic rhinitis.

At least 600 species of fungi are in contact with humans and less than 50 are frequently identified and described in epidemiologic studies on indoor environments (Haleem Khan and Mohan Karuppayil 2012).

Moulds may cause respiratory symptoms as sinusitis similar to the common cold due to inflammation of paranasal sinuses. Mucous membrane irritation syndrome, characterised by symptoms such as rhinorrhoea (running nose), nasal congestion and sore throat, and irritation of nose and eyes, has been found in people exposed to damp buildings (Lanier et al. 2010).

High concentration of moulds, especially *A. alternate,* that develop in very damp environments may be involved in severity of asthma in children and young adults.

Exposure to buildings contaminated with fungi and mycotoxins (*trichothecene*) may develop hypersensitivity pneumonitis, which are a granulomatous lung disease due to exposure and sensitisation to antigens inhaled. This disease can be acute or chronic (Haleem Khan and Mohan Karuppayil 2012).

Exposure to a variety of fungi such as *Aspergillus spp.* and *Fusarium spp.* may result in serious respiratory infections in immunocompromised people. Chronic obstructive pulmonary disease, asthma and cystic fibrosis are disorders among people that are potentially infected with *Aspergillus* (Baxter et al. 2011).

Rheumatic diseases, due to inflammation and stiffness in muscles, joints or fibrous tissue, are exacerbated by environmental conditions, which include dampness, fungi and their products indoors (Breda et al. 2010).

A link between respiratory exposure to fungal material and seasonal allergy was first proposed in 1873 by Blackley, who listed 106 fungi genera including members who elicited allergy (Blackely 1873). The major allergic diseases caused by fungi are allergic asthma, allergic rhinitis, allergic sinusitis, bronchopulmonary mycoses and hypersensitivity pneumonitis (Haleem Khan and Mohan Karuppayil 2012).

References

Adan OCG (1994) On the fungal defacement of interior finishes. Dissertation, Eindhoven University of Technology

Araki A, Kanazawa A, Kawai T et al (2012) The relationship between exposure to microbial volatile organic compound and allergy prevalence in single-family homes. Sci Total Environ 423:18–26. doi:10.1016/j.scitotenv.2012.02.026

Ayerst G (1969) The effects of moisture and temperature on growth and spore germination in some fungi. J Stored Prod Res 5:127–141

Baughman A, Arens E (1996) Indoor Humidity and Human Health–Part I: Literature Review of Health Effects of Humidity- Influenced Indoor Pollutants. ASHRAE Trans 102:193–211

Baxter CCG, Jones AAM, Webb K, Denning DDW (2011) Homogenisation of cystic fibrosis sputum by sonication—an essential step for Aspergillus PCR. J Microbiol Methods 85:75–81. doi:10.1016/j.mimet.2011.01.024

Blackely C (1873) Experimental researches on the causes and nature of *Catarrhus Aestivus* (Hay-Fever and Hay-Asthma).Tindall and Cox, London

Breda L, Nozzi M, De Sanctis S, Chiarelli F (2010) Laboratory tests in the diagnosis and follow-up of pediatric rheumatic diseases: an update. Semin Arthritis Rheum 40:53–72. doi:10.1016/j.semarthrit.2008.12.001

D'Orazio M, Palladini M, Aquilanti L, Clementi F (2009) Experimental evaluation of the growth rate of mould on finishes for indoor housing environments: effects of the 2002/91/EC directive. Build Environ 44:1668–1674. doi:10.1016/j.buildenv.2008.11.004

Daquisto D, Crandell J, Lyons J (2004) Building Moisture and Durability Past, Present and Future Work. U.S. Department of Housing and Urban Development, Washington, D.C

Engvall K, Norrby C, Norbäck D (2002) Ocular, airway, and dermal symptoms related to building dampness and odors in dwellings. Arch Environ Health 57:304–310. doi:10.1080/00039890209601413

Górny RL, Mainelis G, Grinshpun SA et al (2003) Release of *Streptomyces albus* propagules from contaminated surfaces. Environ Res 91:45–53

Grant C, Hunter C, Flannigan B, Bravery AF (1989) The moisture requirements of moulds isolated from domestic dwellings. Int Biodeterior 25:259–284. doi:10.1016/0265-3036(89)90002-X

Haleem Khan AA, Mohan Karuppayil S (2012) Fungal pollution of indoor environments and its management. Saudi J Biol Sci 19:405–426. doi: 10.1016/j.sjbs.2012.06.002

Haseltine E, Rosen J (2009) Dampness and Mould, WHO guidelines for indoor air quality

Hens HLSC (1999) Fungal defacement in buildings: a performance related approach. HVAC and R Res 5:265–280. doi:10.1080/10789669.1999.10391237

IEA (1990) IEA-Annex 14. Condensation and Energy: Guidelines and Practice

Krus M, Sedlbauer K, Zillig W, Kunzel H (2001) A new model for mould prediction and its application on a test roof. In: Proceedings of the 2nd International Scientific Conference on The current problems of Building-Physics in the Rural Building, Cracow, Poland

Künzel H (2011) Building physics and durability–. In: Proceedings of the 11th DBMC International Conference on Durability of Building Materials and Components, Porto, Portugal, pp 1–2

Lanier C, Richard E, Heutte N et al (2010) Airborne molds and mycotoxins associated with handling of corn silage and oilseed cakes in agricultural environment. Atmos Environ 44:1980–1986. doi:10.1016/j.atmosenv.2010.02.040

Moon HJ (2005) Assessing mold risks in buildings under uncertainty. Dissertation, Georgia Institute of Technology

Nielsen KF, Holm G, Uttrup LP, Nielsen PA (2004) Mould growth on building materials under low water activities: influence of humidity and temperature on fungal growth and secondary metabolism. Int Biodeterior Biodegradation 54:325–336. doi: 10.1016/j.ibiod.2004.05.002

Ritschkoff A, Viitanen H, Koskela K (2000) The response of building materials to the mould exposure at different response conditions. In: Proceedings of Healthy Buildings. Espoo, Finland

Sailer MF, Van Nieuwenhuijzen EJ, Knol W (2010) Forming of a functional biofilm on wood surfaces. Ecol Eng 36:163–167. doi:10.1016/j.ecoleng.2009.02.004

Sedlbauer K (2001) Prediction of mould fungus formation on the surface of and inside building components. Report, Fraunhofer Institut for Building Physics

Sedlbauer K, Krus M, Breuer K (2003) Biohygrothermal method for the prediction of mould growth: procedure and health aspects. In: Proceedings 7th International Conference Healthy Buildings. Singapore, pp 666–672

Sedlbauer K, Krus M, Zillig W, Kuenzel H (2001) Mold Growth Prediction by Computational Simulation. ASHRAE Conference IAQ. San Francisco, USA

Shirakawa MA, Gaylarde CC, Gaylarde PM et al (2002) Fungal colonization and succession on newly painted buildings and the effect of biocide. FEMS microbiol ecol 39:165–73. doi:10.1111/j.1574-6941.2002.tb00918.x

Shirakawa M a., Loh K, John VM, et al. (2011) Biodeterioration of painted mortar surfaces in tropical urban and coastal situations: Comparison of four paint formulations. International Biodeterioration & Biodegradation 65:669–674. doi: 10.1016/j.ibiod.2011.03.004

Tamburlini G, Von Ehrenstein OS, Bertollini R (2002) Children' s health and environment: A review of evidence

Vacher S, Hernandez C, Bärtschi C, Poussereau N (2010) Impact of paint and wall-paper on mould growth on plasterboards and aluminum. Build Environ 45:916–921. doi:10.1016/j.buildenv.2009.09.011

Viitanen H, Hanhijarvi A, Hukka A, Koskela K (2000) Modeling mould growth and decay damages. In: Proceedings of Healthy Buildings, Espoo, Finland, pp 341–346

Viitanen H, Ojanen T, Peuhkuri R et al (2011) Mould growth modelling to evaluate durability of materials. In: Proceedings of the 12th DBMC International Conference on Durability of Building Materials and Components, Porto, Portugal

Chapter 5
Analytical and Experimental Methods for the Assessment of the Biological Proliferation in Buildings

Abstract In order to preserve buildings from the colonisation of microorganisms and to act efficiently against biodeterioration, it is necessary to have a better understanding of biodeterioration mechanisms and their effects on materials properties. Consequently, there is a growing demand for calculation methods in building engineering to assess the moisture behaviour of building components and microorganism risk prediction in order to ensure a healthy environment and to avoid defacement of materials and other social and economic consequences. Many building hygrothermal analysis methods are able to simulate the coupled transport processes of heat and moisture for one or multidimensional cases, aiming to predict biological risk. Additional measurements in laboratory and in situ conditions have been used for the validation of these models. In the first part of this chapter, we review some of the major biological risk predictive models, both inside and outside the buildings. Then, in the second part, we will describe some methods of accelerated experimental testing for the evaluation of biological defacement of building materials. A more in-depth study of microorganism growth under transient conditions is still necessary in order to define the most reliable prediction model. To do so, additional measurements in laboratory and in situ conditions on new Nearly Zero Energy Buildings and components would be desirable.

Keywords Condensation · Time of wetness · VTT model · Isopleth · Biohygrothermal model · Accelerated test · Biodeterioration

5.1 Analytical Models

In literature there are many hygrothermal analysis methods to simulate the coupled transport processes of heat and moisture for one or multidimensional cases.

Hundreds of building software tools based on these methods have been developed or enhanced to be used for the prediction of the hygrothermal performance of buildings. These models vary significantly concerning their

E. Di Giuseppe, *Nearly Zero Energy Buildings and Proliferation of Microorganisms*,
SpringerBriefs in Applied Sciences and Technology, DOI: 10.1007/978-3-319-02356-4_5,
© The Author(s) 2013

mathematical sophistication that depends on the degree that takes into consideration the following parameters: moisture transfer dimension; type of flow (steady-state, quasi-static, or dynamic); quality and availability of information and stochastic nature of each data (material properties, weather, construction quality, etc.) (Delgado et al. 2010).

As the purpose of most hygrothermal models is usually to provide sufficient and appropriate information needed for decision-making, the software should be available in the public domain (freeware or commercially) and it should be "user friendly" (Delgado et al. 2010).

In the following sections, we will see some of the most known models for the evaluation of the development of microorganisms on the facades and inside of buildings.

5.1.1 Models of Biological Growth on Facades

Many studies have pointed out that microbiological growth on building facades is due to the high values of surface moisture content, which results from the combined effect of four parameters: surface condensation, wind-driven rain (WDR), drying process and properties of the exterior layer.

Most analytical methods of the envelope hygrothermal performance are developed in this way: they are useful tools in assessing exterior condensation on façades and the importance of radiative balance on the exterior surface temperature. The periods of surface condensation and the accumulated degree of cooling below dew point temperature are taken as criterion to classify the resulting biological growth (Krus et al. 2006).

However, no simple method has yet been developed to predict the risk of the biological defacement of the façade similarly to what has already been done for mould growing on interior finishes, where the mould spore hygrothermal behaviour is taken into account.

Surface humidity is considered as the principal criterion for assessing the risk of biological growth. Nevertheless, the comparison of simulated values with the results of ''in situ'' tests performed on a façade covered with ETICS showed that there is no good agreement between the simulated and the measured values of the relative humidity, especially when wind-driven rain is taken into account (Delgado et al. 2010).

Recent studies have started to develop a simple process to predict the risk of biological defacement of building facades, also by analysing experimental data on growth.

An interesting attempt is the model "BIO.MOD", which defines three risk indices (Eqs. 5.1, 5.2 and 5.3) that are related to the surface humidification, by condensation (CPE_a), due to WDR ($WDRPEa$) or due to the sum of the two, with the maximum drying capacity (DPE_a) (Barreira and Freitas 2011):

$$BIO.MOD1 = \frac{CPE_a}{DPE_a} * 10^3 \left[\frac{Pa * h}{Pa * h} \right] \tag{5.1}$$

$$BIO.MOD2 = \frac{WDRPE_a}{DPE_a} * 10^3 \left[\frac{kg/m^2}{Pa * h} \right] \tag{5.2}$$

$$BIO.MOD3 = \frac{CPE_a + WDRPE_a}{DPE_a} * 10^3 \left[\frac{Pa * h + kg/m^2}{Pa * h} \right] \tag{5.3}$$

For Barreira and Freitas, exterior surface condensation can be analysed using psychrometry principles (Barreira and Freitas 2011). When water vapour partial pressure of the air is greater than the water vapour saturation pressure on the surface, condensation will occur (Hagentoft 2001). According to Zheng et al., the difference between the water vapour partial pressure in the air (Pv(air), in Pa) and the water vapour saturation pressure on the surface (Psat(surface), in Pa) may be called Condensation Potential (CP, in Pa), which implies condensation for positive values (Zheng et al. 2004). CP can be understood as the amount of water vapour that is available to condensate (Eq. 5.4).

$$CP = P_v(air) - P_{sat}(surface) \tag{5.4}$$

To evaluate the amount of condensation, positive CP and its lasted time should be considered. The product of positive CP (CP (>0), in Pa) by its lasted time (Δt_{CP} (>0), in h) may be called Condensation Potential Equivalent (CPE, in Pa) and allows the estimation of the risk of condensation for a certain period of time (Eq. 5.5). In order to estimate the risk of condensation for a certain period of time CPE must be accumulated in time (CPEa).

$$CPE = CP_{(>0)} * \Delta t_{CP_{(>0)}} \tag{5.5}$$

The humidification of a façade due to wind-driven rain (WDR) may be assessed, for a certain period of time, through the WDR Potential Equivalent (WDRPEa, in kg/m^2), which is obtained by integrating the intensity of WDR (in $kg/(m^2 s)$) in time. WDRPEa has to be multiplied by 100 in order to reach values that are comparable with CPEa values (Eq. 5.6).

$$WDRPE_a = 100 * \int_0^t WDRdt \tag{5.6}$$

Similar to condensation, also the drying capacity of a wet surface can be analysed using psychrometry principles (Hagentoft 2001). By analogy, it is possible to establish the concept of Drying Potential (DP, in Pa), as being the difference between the water vapour saturation pressure on the surface (Psat(surface), in Pa) and the water vapour partial pressure in the air (Pv(air), in Pa), which implies evaporation for positive values (Eq. 5.7). DP can be understood as the amount of water vapour transferred to the air, considering that the surface remains permanently wet.

$$DP = P_{sat}(surface) - P_v(air) \tag{5.7}$$

In order to evaluate the maximum ability to dry out, the product of positive DP (DP(>0), in Pa) by its lasted time (ΔtDP(>0), in h) shall be considered and may be called Drying Potential Equivalent (DPE, in Pa) (Eq. 5.8). In order to estimate this ability for a certain period, DPE must be accumulated in time (DPEa).

$$DPE = DP_{(>0)} * \Delta t_{DP_{(>0)}} \tag{5.8}$$

It must be stated that DPEa is not useful as a parameter for modelling the real drying capacity of a wet surface, as this is not permanently saturated. After some time, the liquid water evaporates and the vapour pressure at the surface depends not only on the surface temperature, but also on its relative humidity. However, in order to avoid the use of relative humidity and to simplify the parameters used in the drying process assessment, DPEa can be employed as an overvalued drying capacity.

Using the data collected during an "in situ" campaign, Barreira and Freitas calculated in annual bases CPEa, WDRPEa, DPEa and the three indices of BIO.MOD (Barreira and Freitas 2011). They found a good agreement between the index BIO.MOD3, which combines surface condensation with the effect of WDR and "quantifies" the risk of defacement due to biological growth, and the accumulated hours of surface saturation (relative humidity equal to 100 %), measured simultaneously. Using the model BIO.MOD they also affirmed that the drying process is the most relevant parameter and that surface condensation has more impact than WDR. The authors finally establish a risk map for walls covered with ETICS located in the Portuguese territory.

The development of further models, which take into account the types of microorganisms proliferating, still requires more research and analysis of experimental data.

5.1.2 Models of Internal Mould Growth

In order to better understand the phenomena of formation of mould on the construction elements, dynamic models are needed, which are able to consider not only the variations of the internal and external boundary conditions of relative humidity and temperature, but also the time required for the growth of mould. These are the cardinal influencing factors in the mould growth process (Adan 1994). The critical value for these parameters can however differ for each mould species.

While in the past the temperature ratio was often used to minimise the risk of mould, nowadays more advanced mould prediction models can be found. These models include the main influencing factors for mould growth, which are surface temperature and relative humidity.

Next sections will give an overview of the different existing models on the mould risk evaluation.

5.1.2.1 IEA-Annex 14

IEA-Annex 14 (IEA 1991; IEA 1990) stated that:

- Surface condensation starts each time the relative humidity (RH) on a surface reaches 100 %, that means, each time the vapour pressure (*P*) in the air against the surface equals or becomes higher than the saturation pressure on the surface (P'si):
  Surface condensation when P ≥ P'si
- Mould germination becomes possible when the mean water activity against/on a nutrient surface remains higher than a threshold value 'a', 'a' being a function of the mould species, the temperature, the substrate (nutrient), etc.... Using the fact that, in steady-state, the water activity is nothing other than the RH, the mould condition becomes:
  Mould germination when P ≥ a*P'si

IEA-Annex then defined a surface relative humidity threshold for mould growth dependent on the elapsed time, based on the lowest isopleth for *Aspergillus versicolor*: 80, 89 or 100 % for an exposure time of 1 month, 1 week and 1 day, respectively.

The Annex also defined a design value for the temperature ratio (Eq. 5.9):

$$\tau = \frac{\theta_{s,\min} - \theta_e}{\theta_i - \theta_e} \geq 0.7 \tag{5.9}$$

With $\theta_{s,\min}$(°C) being the minimum indoor surface temperature and θ_i and θ_e the inside and outside temperature (°C), respectively. A temperature ratio of 0.7 is proposed as criterion, related to an acceptable mould risk of 5 %. A lower ratio introduces an unacceptable high mould risk.

Although the temperature ratio is often used as a design criterion, an in situ study on 35 mould infested dwellings indicated that this criterion may not be used as a stand-alone performance because of the importance of other factors such as low ventilation rate, rain infiltration, less heating or thermal bridges (Hens 2003).

Furthermore, a relative humidity threshold of 100 % in cases of 1-day exposure is questionable since liquid water hinders mould development. A threshold of 99 % relative humidity is more plausible (Vereecken and Roels 2012).

5.1.2.2 Model of Time of Wetness

Adan studied fungal growth not only under steady-state but also under transient indoor conditions, in order to improve the understanding of the process which induces the fungal defacement of interior finishes (Adan 1994). He particularly investigated the response of the fungal cell to transient water vapour pressures, and observed that fungi are capable of an instantaneous water vapour uptake as the RH increases, suggesting that short periods of high RH should not be neglected in the evaluation of indoor climate with respect to mould growth.

In order to indicate the water availability under transient conditions, he introduced the time of wetness (TOW), defined by the ratio of the cyclic wet period (i.e. when the RH $\geq$ 80 %) and the cyclic period (Eq. 5.10):

$$TOW = \frac{\text{cyclic wet period } (RH \geq 80\%)}{\text{cyclic period } (wet + dry)} \qquad (5.10)$$

His preliminary experiments indicated that the growth of *Penicillium chrysogenum* on gypsum-based finishes is only weakly affected for a TOW $\leq$0.5. The RH value during the drying periods hardly shows any influences on the fungal growth-TOW relation. Furthermore, except for very fast oscillations of the RH, the frequency of high RH periods only slightly affects the TOW effects on fungal growth.

Besides, Adan developed sigmoid curves, which enable a satisfactory fit of the mould growth rate of *P. chrysogenum* on gypsum board material. In this, the average rating is defined as in the BS3900 standard (Table 5.1).

By the way, Adan's experiments were limited to measured data on gypsum board materials inoculated with *P. chrysogenum*, so his model cannot be used to predict fungal growth in cases of other substrates or species.

5.1.2.3 VTT Model

The VTT model is an empirical mould prediction model developed by Hukka and Viitanen (1999), in which quantification of mould growth is based on the mould index (M) used in the experiments on wooden materials for visual inspection (already seen in Table 4.1). The index can be used as a design criterion, e.g. often

Table 5.1 Rating scale according to mould coverage area in the model of Adan (1994)

Rating	Coverage area
0	No mould growth
1	Coverage $\leq$1 %
2	1 %$\leq$ Coverage $\leq$10 %
3	10 %$\leq$ Coverage $\leq$30 %
4	30 %$\leq$ Coverage $\leq$70 %
5	70 %$\leq$ Coverage

a mould index equal to 1 is defined as the maximum tolerable value since from that moment the germination process is assumed to start.

With the exception of the maximum index of 6, the definition of the mould indices agrees very well with the definition for the rating used by Adan (Table 5.1).

The model consists of differential equations describing the growth rate of the mould index in different fluctuating conditions, including the effect of exposure time, temperature, relative humidity and dry periods.

The incremental mould index can be calculated by using a differential equation.

For pine and spruce, the incremental change in mould index (M) is given by Eq. 5.11:

$$\frac{dM}{dt} = \frac{1}{7\exp(-0.68\ln T - 13.9\ln RH + 0.14W - 0.33SQ + 66.02)} k_1 k_2 \quad (5.11)$$

where T (0,1–40 °C) is the ambient temperature, RH (%) is the Relative Humidity, W the wood specie (0 for pine, 1 for spruce), SQ the surface quality (0 for resawn and 1 for original kiln-dried timber), the factor k_1 defines the growth rate under favourable conditions, and k_2 represents the response time for the beginning of mould growth.

The main disadvantage of the first VTT model is its limitation to unpolluted spruce and pine softwood. This type of wood is widely used in Scandinavia, but hardly used for building construction elsewhere.

Recently, the VTT model was expanded for other building materials (Ojanen et al. 2010): spruce board (with glued edges), concrete (K30, maximum grain size 8 mm), aerated concrete, cellular concrete, polyurethane thermal insulation (PUR, with paper surface and with polished surface), glass wool, polyester wool and expanded polystyrene (EPS). Pine sapwood was used as a reference material.

The mould growth parameter values of different materials were adapted to the existing model. Some improvements were applied for the model structure to better adjust different growth phenomena.

The main difference with the original model was found at microscopic level, since for some materials already at this level a rather high mould growth coverage could be observed. Therefore, the mould index determination was updated with these microscopic growth coverage findings in index levels three and four.

The principle when updating the original mould growth model for other materials was that new values for the factors presented in Eq. 5.11 were determined for these materials using the results from several experiments. To improve the usability of these new values, they were not presented as exact values for each material but as "material classes" according to the sensitivity of the material to mould growth. Four different sensitivity classes were determined (Table 5.2).

The presented improvements of the mould growth simulation model do not guarantee the exact prediction of mould in all cases and conditions. The new model values were determined based on a limited set of experiments with relatively large scattering and should therefore be interpreted as a first approximation.

Table 5.2 Mould sensitivity classes (Vereecken and Roels 2012)

Sensitivity class	Material in experiment	Material groups
Very sensitive	Pine sapwood	Untreated wood; includes lots of nutrients for biological growth
Sensitive	Glued wooden boards, PUR with paper surface, spruce	Planed wood; paper-coated products, wood-based boards
Medium resistant	Concrete, aerated and cellular concrete, glass wool, polyester wool	Cement or plastic based materials, mineral fibres
Resistant	PUR with polished surface	Glass and metal products, materials with effective protective compound treatments

The variation of the material sensitivities is high, estimation of a product sensitivity class is difficult without testing, the surface treatments may enhance or reduce growth potential, different mould species have different requirements for growth, and the evaluation of the actual conditions in the critical material layers may include uncertainties.

5.1.2.4 Isopleth Models

Isopleth curves represent the relation between the mould risk and the main mould inducing factors (relative humidity or water activity, temperature, exposure time). These curves separate favourable from unfavourable conditions for mould growth. The simplest models just provide the limit state curve, more advanced isopleth models subdivide in time until germination and growth rate.

In the model developed by Clarke et al. the mould fungi found in buildings are subdivided into six categories, ranging from xerophilic (dry loving) to hydrophilic (wet loving) fungi (Clarke et al. 1997, 1999). Each category constitutes a family of mould species possessing similar growth requirements over the range of temperature and humidity conditions likely to be found in the indoor environment. Each category is accompanied with a representative fungus:

- Highly xerophilic: *Aspergillus repens*
- Xerophilic: *A. versicolor*
- Moderately xerophilic: *P. chrysogenum*
- Moderately hydrophilic: *Cladosporium sphaerospermum*
- Hydrophilic: *Ulocladium consortiale*
- High hydrophilic: *Stachybotrys atra*.

For each of these categories a growth limit curve defined by a third-order polynomial function was determined based on an analysis of published data. When the relative humidity and temperature combination exceeds such a curve, mould growth of the matching fungi will occur (Fig. 5.1).

Fig. 5.1 Limiting growth curves for six mould growth categories. Reprinted from Clarke et al. (1999), copyright 1999, with permission from Elsevier

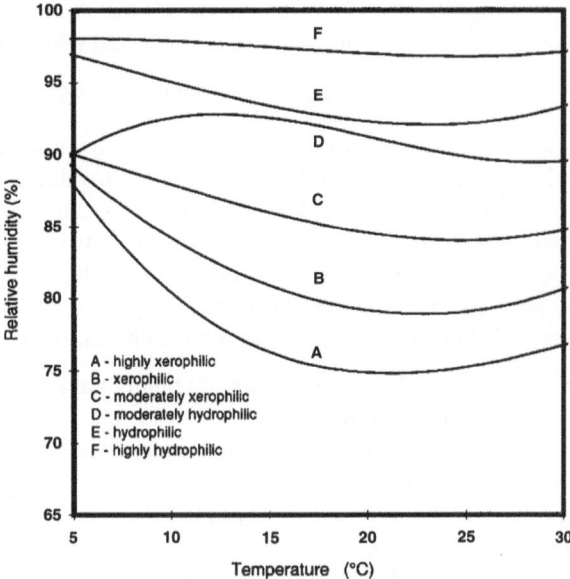

The mould growth prediction model was developed in the ESP-r modelling system for the assessment of the environmental and energy performance of buildings. This model is able to predict the time series evolution of the local surface temperature and relative humidity while taking explicit account of constructional moisture flow and local air movement. The mould growth limit curves are contained in a moulds database. This allows the predicted local conditions to be plotted directly to the mould growth curves. The concentration of plotted points relative to the various growth bands allows an assessment to be made of the risk of mould growth and its persistence over time.

Because the mould growth curves were generated from experiments in which the principal determinants of growth were maintained at fixed values, the model is unable to indicate the effects on mould growth of temperature and/or humidity fluctuations over prolonged periods of time.

Since in the model the exposure time is not taken into account, a single excess of the isopleths is set equal to mould formation, consequently this model assumes the worst-case scenario and risks to overestimate the mould risk on finishing materials.

Among the moulds frequently found on defaced finishes in buildings, the xerophilic species *A. versicolor* has the isopleth with the lowest relative humidity for germination (IEA 1991). This lowest isopleth is closely matched by the following parabolic relation (the relative humidity RH in %, temperature θ in °C) (Hens 1999) (Eq. 5.12):

$$RH_{threshold} = 0.033\theta^2 - 1.5\theta + 96 \tag{5.12}$$

Giving 79 % at 22.7 °C. At 10 °C, the threshold raises to 84 %. For shorter periods, a logarithmic function is suggested (Eq. 5.13):

$$RH_{threshold} = min\{1, 0.8[1.25 - 0.075 \ln(t)]\} \tag{5.13}$$

where t is the time (days). This function developed for shorter periods does not take into account the influence of the temperature.

Because of the abundance of mould species and materials, an individual isopleth system for each species and substrate is not possible. Therefore, in order to expand the existent isopleth system to date, Sedlbauer subdivided the mould species and materials found in buildings in a set of classes (Sedlbauer 2002). A first subdivision was based on the health risk of the different mould species:

• Class A: mould species which are highly pathogen and consequently not allowed to occur in buildings;
• Class B: mould species which are pathogen when exposed over a longer period or which cause allergic reactions;
• Class C: mould species which are not dangerous to health.

Based on measurements for a collection of species, minimum growth conditions for the three hazardous classes were obtained. For classes B and C only slightly different results were obtained, so that classes B and C were combined in one class B/C.

The conditions below where no spore germination or growth will occur was indicated by the LIM (Lowest Isopleth for Mould)-curve. This was developed based on the lowest envelope of all the lowest curves of the group.

After the determination of the LIM-curves, representative mould fungi for the different hazardous classes were searched. They had a LIM-curve which approximated the LIM-curve described above. For class A this representative fungus is *A. versicolor*, for class B/C the mould fungi are *Aspergillus amstelodami, Aspergillus candidus, Aspergillus ruber* and *Wallemia sebi*. Based on the growth of these representative fungi on an optimal culture medium, the isopleth systems for the hazardous classes were developed.

The influence of building substrate, i.e., the building material itself and its possible soiling, was then taken into account by a second subdivision in the following categories (Sedlbauer 2002):

Substrate category 0—Optimal culture medium;

Substrate category I—Biologically recyclable building materials like wall paper, plaster cardboard, building materials made of biologically degradable raw materials, material for permanent elastic joints;

Substrate category II—Biologically adverse recyclable building materials such as renderings, mineral building material, certain wood as well as insulation material not covered by I;

Substrate category III—Building materials that are neither degradable nor contains nutrients.

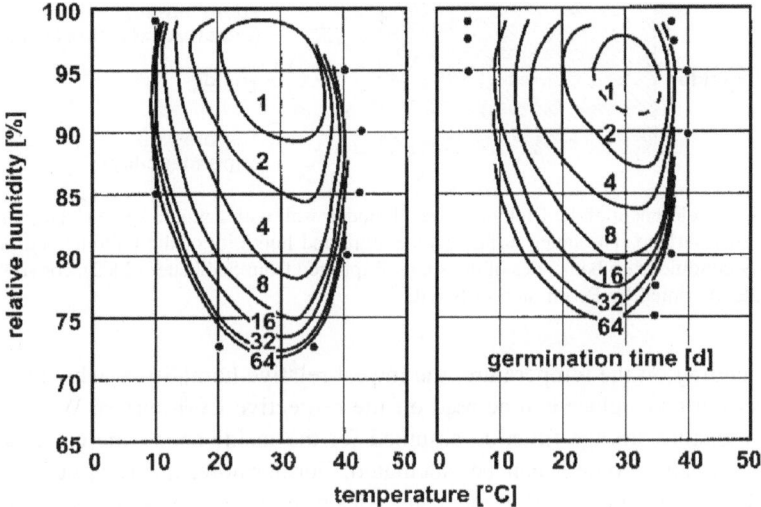

Fig. 5.2 Isopleths for mould spores of *Aspergillus restrictus* (*left* side) and *A. Versicolor* (*right* side) Reprinted from Sedlbauer (2002), copyright 2002 by SAGE, reprinted by Permission of SAGE

Substrate based isopleths are made in such a way that always the worst-case scenario is examined. For the mould growth on building materials the study considered a germination graph and a growth rate graph (Fig. 5.2).

5.1.2.5 Biohygrothermal Model

To make a more reliable prediction of the mould risk possible in cases of transient conditions, Sedlbauer extended his isopleth model with the Biohygrothermal model (Sedlbauer 2002). This model makes it possible to calculate the moisture balance of a spore independently of the transient boundary conditions, thus letting us consider interim drying out of the fungus spores (Fig. 5.3).

The Biohygrothermal model for predicting the germination of the spores is based on the fundamental idea that a fungus spore has a certain osmotic potential, so that spores can absorb water existing in the environment, i.e. in materials as well as in the air.

This potential is computationally described by means of a moisture retention curve. The absorption of humidity through the spore septum is described by diffusion, until certain moisture content inside the spore is reached that is needed for starting the metabolism. From this point on the fungus can regulate its metabolism, if necessary even independently of the surrounding conditions.

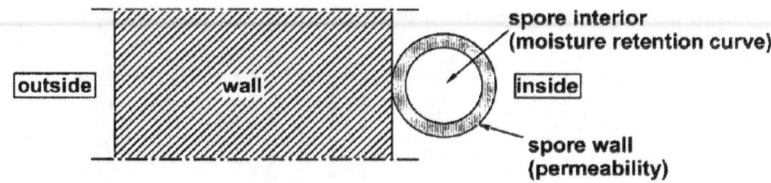

Fig. 5.3 Development of the Biohygrothermal model: wall with a mould spore (*highly enlarged*) on the inner surface. The inner surface temperature and humidity of the building wall serve as boundary conditions on both sides of the spore. Reprinted from Sedlbauer (2002), copyright 2002 by SAGE, reprinted by Permission of SAGE

Depending on the temperature, the lowest relative humidity at which the spore germination takes place can be read off the respective LIM-curves. With the help of the moisture storage function assumed for the inside spore, the corresponding critical moisture content can be calculated. Furthermore, the LIM-curves in the isopleth systems of the appropriate categories of building materials have to be used when setting the critical moisture content.

The Biohygrothermal model has been implemented in the Wufi[1] software, which includes diffusion, liquid transport and moisture storage processes.

5.2 Accelerated Experimental Testing

To preserve buildings from the colonisation of microorganisms and to act efficiently against biodeterioration, it is necessary to have a better understanding of biodeterioration mechanisms and their effects on the properties of materials. Several tests to study biodeterioration of building materials exist. Among them some were developed without accelerated weathering of the matrix leading to experiment time ranging from several month to some years, while more recently several studies have focused on the assessment of biodeterioration of building materials through laboratory accelerated ageing tests.

Laboratory tests have to be as realistic as possible. They are required to be "accelerated", reproducible, low cost and easy to implement in construction material laboratories. They also have to discriminate among the support parameters for biological growth (Escadeillas et al. 2006). The concerned topic is very broad, not only for the wide variety of materials on the market but also for the range of microorganisms.

Some examples of interesting test methods developed to date are provided in the following sections.

---

[1] WUFI® (Wärme und Feuchte instationär) is a software family, which allows realistic calculation of the transient coupled one- and two-dimensional heat and moisture transport in multi-layer building components exposed to natural weather.

5.2.1 Algae

Dubosc developed a general methodology for ageing tests on mortar samples (species selection and preparation, sample preparation, quantification techniques, etc...), and tested several environmental conditions which led to the design of complementary tests using different moistening modes (capillary sorption, water flow, spraying, different relative humidity levels) (Dubosc 2000). Validation tests performed on different mortar mixtures showed that some types of conditions give positive results within a 2 months period and may be used for further studies. Tests also pointed out that material porosity and roughness are very influencing factors concerning biological growth on cementitious support.

Later, together with his research group, Dubosc further developed two test methods (Escadeillas et al. 2006):

- A static test simulating growth conditions at the base of a construction. This method simulated algal growth at a wall base, where water supplies occurred by capillary ascent. It highlighted the influence of the material parameters such as the mineral composition and the pore-size distribution in the mortars. Here, algae were tested individually (one algal species per box).
- A dynamic test simulating run-off on some parts of constructions (this test corresponds to external wall surfaces exposed to rain, or leaky parts of a building or design defects). In this test, the mortar prisms were tilted at 45° in a polycarbonate transparent chamber. The faces to be studied were subjected to intermittently applied, uniform run-off of a solution inoculated with a mixture of three algae (Fig. 5.4).

Characterisation tests were performed on mortars. This choice was a simplification with respect to the initial study, which concerned concrete external walls, but in practice, the assimilation of the first few millimetres of a concrete wall to a mortar is not absurd as regards wall effect phenomena. Moreover, the use of mortars allowed small sized specimens to be made. Results showed that the algal growth between 45 and 100 days was considerable and differed according to the mortar support (Fig. 5.5).

The authors then proposed some technical methods to characterise colonised areas (Escadeillas et al. 2009). The first non-destructive method allowed a covered surface area to be estimated by image analysis (based on a clustering method). The second method, also non-destructive, allowed the density of the algae, which directly influences the stain intensity, to be estimated by spectrophotometry (or colorimetry). The third method, which is destructive and is based on the measurement of chlorophyll, was used to make a comparative quantification of the algae on various colonised mortars and to obtain information on the algal state of vitality.

Barberousse et al. used accelerated water-streaming test methods to evaluate facade coatings prone to colonisation by algae and cyanobacteria, by closely

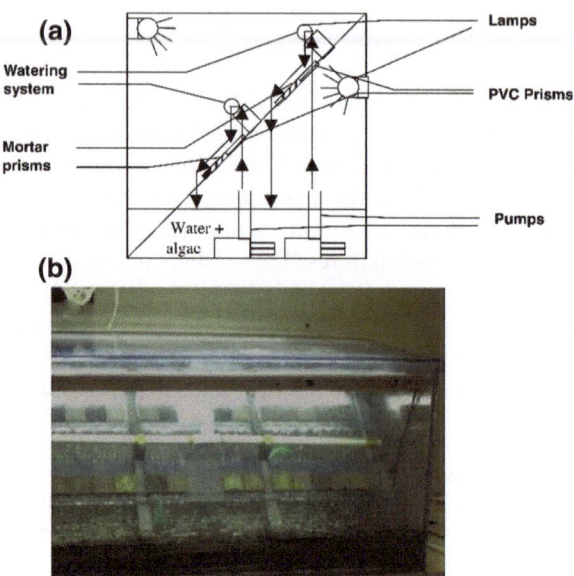

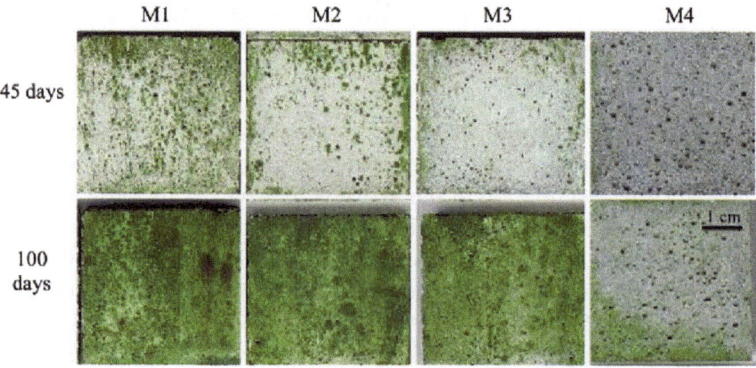

Fig. 5.4 a Run-off test schematic diagram. **b** Run-off test picture. Reprinted from Escadeillas et al. (2006), copyright 2006, with kind permission from Springer Science and Business Media

Fig. 5.5 Macroscopic aspect of mortar prisms after 45 and 100 days. Reprinted from Escadeillas et al. (2009), copyright 2009, with kind permission from Springer Science and Business Media

reproducing the phenomenon of natural biological soiling (Barberousse et al. 2007).

The water-streaming test mimics those conditions of colonisation by producing the stream of a culture of microorganisms on top of materials. The device was similar to the one previously used to investigate concrete materials (Dubosc 2000), with some improvements to investigate parameters of materials favouring algal growth.

Fig. 5.6 Setup of equipment used by Barberousse et al. to verify the colonisation of the facades by microalgae and cyanobacteria. Reprinted from Barberousse et al. (2007), copyright 2007, with permission from Elsevier

The system consisted of a $100 \times 50 \times 50$ cm glass chamber containing stainless steel supports inclined at $45°$ onto which specimens of facade coatings were placed (Fig. 5.6). The chamber was filled with 50 l of BBM (Bold's Basal Medium) enriched with algae or cyanobacteria cultures.

The device was equipped with two sprinkling rails made of stainless steel tubes with 2 mm diameter holes drilled at every centimetre. The rails were supplied by pumps immersed in the suspension and connected to the rails.

Thus, the principle of the water-streaming test is as follows: the suspension circulating through the sprinkling rails was directed onto the top of the specimens and ran down their surface, allowing algal and cyanobacterial cells to adhere to the surface of the specimens, depending on their characteristics, as they would in nature. The suspension was then recycled by the pumps and sprayed again onto the specimens.

The sprinkling cycles were set to start every 12 h and to run for 90 min; the amount of suspension received by each specimen of material during a cycle was 20 ± 2 lh^{-1}. Furthermore, since algae and cyanobacteria need light to grow, the setup also included two 30 W neon lamps placed at the same distance from the centre of the chamber: the illumination received by each specimen lasted a day length of 12 h and was set to start with the beginning of a sprinkling cycle. The glass chamber was placed in a dark room and conditioned at 23 ± 2 °C and 50 ± 5 % relative humidity.

Barberousse et al. then evaluated the colonisation kinetics of materials by image analysis. The surface of each specimen exposed to colonisation was digitised weekly using an office scanner. The obtained numerical image was treated to establish a histogram of the number of pixels versus their intensity.

The colonisation process results obtained reproduce algal growth often observed on building envelopes. This confirms that the principle of the device, by wetting the sloped materials, mimicked the humidification of facades by liquid water, with the difference that in accelerated tests the materials are inclined and the sprinkling solution flow is high in order to accelerate the colonisation process. Porosity and roughness of the materials showed to be parameters of great influence on algae and cyanobacterial establishment.

De Muynck et al. designed a modular setup for accelerated water run-off tests, which allowed both simultaneous and separate evaluation of 12 different surface treatments for preventing algal fouling on white architectural and autoclaved aerated concrete (De Muynck et al. 2009). The modular setup consisted of 12 stainless steel compartments supported by a wooden frame at 45° inclination (Fig. 5.7).

Each compartment was equipped with a sprinkling rail on top and a plastic gutter at the bottom. A transparent 2 l PET bottle beneath each compartment served as reservoir for the algal cultures. Circulation of the algal cultures was achieved by means of an aquarium pump (200 lh^{-1}) immersed in the PET bottle. Algal cultures were introduced at the top of the compartment by means of a plastic tube connected to the sprinkling rail. Water running down from the specimens was subsequently collected by means of the gutter and a funnel covering the PET bottle.

The run-off period was set to start every 12 h and ran for 90 min. Furthermore, the setup was submitted to a 12 h day and night regime, which started simultaneously with the run-off periods. During the day regime, light was provided by means of 30 W lamps. The temperature and relative humidity ranged between 19.5 °C (night) −21.5 °C (day) and 86 % (day) −93 % (night), respectively.

Every week, the contents of the reservoirs were replaced by new algal cultures, after cleaning of the reservoirs. Additionally, every 2 weeks, the reservoirs were replaced by new ones.

The use of a modular setup also allowed testing of several biocidal treatments at the same time. In this way, leaching of biocidal compounds did not have any influence on the performance of other treatments as would be the case with

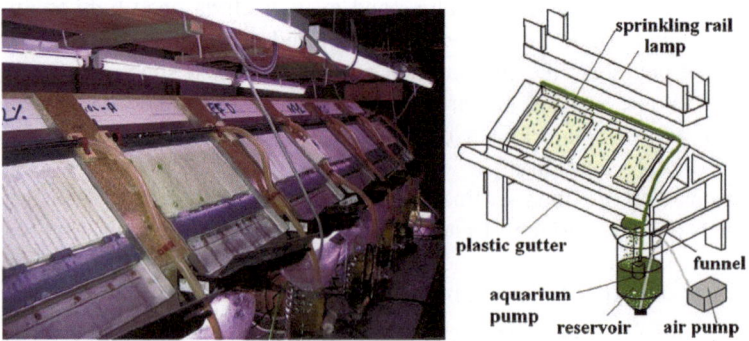

Fig. 5.7 Modular setup used for the accelerated fouling of building materials by means of algae. The image on the right gives a schematic presentation of 1 unit. Reprinted from De Muynck et al. (2009), copyright 2009, with permission from Elsevier

previous setup (Barberousse et al. 2007), where only one algal reservoir was used for all the specimens belonging to different treatments.

Furthermore, due to the weekly replacement of the algal cultures, the contribution of leached compounds on the overall performance of the treatment was greatly decreased.

Algal fouling on samples was evaluated by means of colorimetric and image analysis, for which the authors proposed new evaluation criteria, based upon a colour threshold in the CIELab[2] colour space (Fig. 5.8).

Results showed that for white concrete, contrary to untreated specimens which had 40 % of the surface covered with algae, no fouling was observed for surface treated specimens after 12 weeks of exposure to algae under test conditions. For autoclaved aerated concrete (AAC), the different strategies examined were unable to completely prevent the algal fouling. The use of water repellents resulted in green algal streaks along the surface. Biocide treated specimens showed a delay of onset of fouling of 2–4 weeks under the test conditions. Combinations of water repellents and biocides appear to be the most effective treatments for the prevention of algal fouling on concrete with a high bioreceptivity.

5.2.2 Mould

Shirakawa et al. developed and standardised an accelerated laboratory test for detecting bioreceptivity of indoor mortar to fungal growth (Shirakawa et al. 2003).

To determine which fungal species were predominant under field conditions, they used mortar samples collected from 41 buildings in two cities of Sao Paulo State in the South East of Brazil.

Then four different mortars, two laboratory-manufactured mortars composed of ordinary Portland cement, high calcium hydrated lime and standardised sand, and two different ready-mixed building mortars from the Brazilian market, were investigated for their susceptibility to colonisation by *C. sphaerospermum.*

Each of the tested mortar samples was aseptically inoculated with 25 µl of a spore suspension by placing a droplet in the centre of mortar specimen.

Large mortar samples were exposed to three different RH levels (75, 85 and 100 %), generated using saturated solutions of NaCl, KCl and pure water, respectively. Tests at 85 and 100 % RH were carried out with fungal inocula. Each mortar specimen was set in a tightly closed glass flask (600 cm^3). After inoculation, samples were incubated at 25 °C for 30 days. Interaction of *C. sphaerospermum* with mortar specimens was studied using techniques of scanning and

---

[2] A Lab colour space is a colour-opponent space with dimension L for lightness and a and b for the colour-opponent dimensions, based on nonlinearly compressed CIE XYZ colour space coordinates. CIE coordinates are based on a cube root transformation of the colour data.

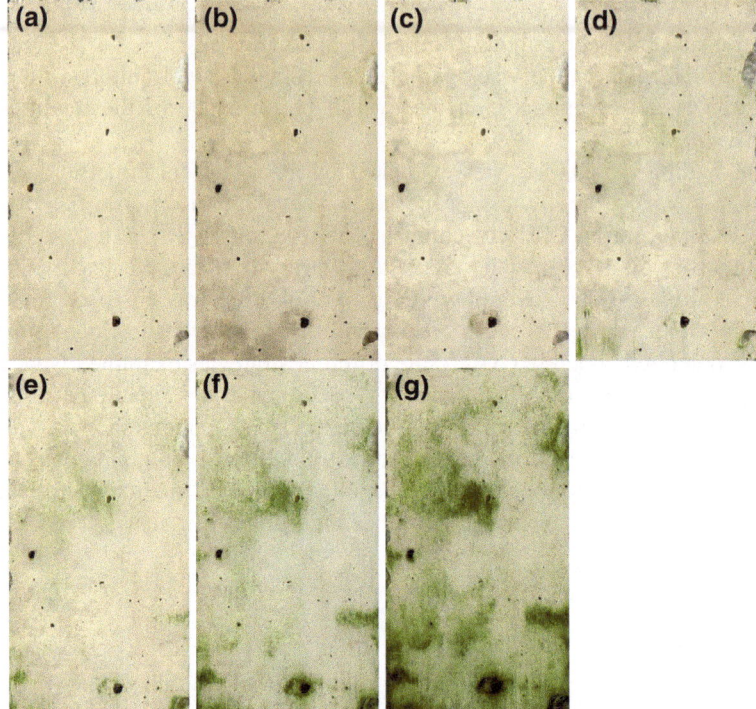

Fig. 5.8 Evolution of the visual appearance of the untreated white concrete specimens after 0 (**a**), 2 (**b**), 4 (**c**), 6 (**d**), 8 (**e**), 10 (**f**) and 12 (**g**) weeks of accelerated fouling tests. Reprinted from Muynck et al. (2009), copyright 2009, with permission from Elsevier

environmental scanning electron microscopy combined with energy dispersive X-ray analysis.

The application of the test demonstrated differences in the bioreceptivity of the four types of mortars, as revealed by light and electron microscopy studies. Parameters such as the type of substratum for casting mortars, the size of test specimens, degree of mortar carbonation and relative humidity to which mortar samples were exposed, proved to be key factors influencing the fungal bioreceptivity of mortars.

Also Urzì and De Leo carried out experiments in laboratory conditions, as well as outdoors, with artificially infected mortars to study the effectiveness of water repellents and biocides (Urzì and De Leo 2007). In the first case, the efficiency of treatments was tested against a massive colonisation of different kinds of microorganisms (bacteria, fungi and a mixture containing cyanobacteria, algae, bacteria and fungi). In the second set of experiments, under outdoor conditions, the authors observed the natural settlement of airborne microbiota on untreated and treated mortars, exposed in the city of Messina (Italy), facing North East and kept inclined at an angle of 45°.

During laboratory tests, untreated and treated mortar probes were inoculated in duplicate with fungal, bacterial and algal suspensions separately, and maintained in constant humid condition at room temperature and day light.

The progression of microbial colonisation was monitored through stereomicroscopic observations at intervals of 1 month. Fifteen months after the inoculation, all the experiments were stopped and one replica of each mortar probe was utilised for light and epifluorescent microscopic analysis and for microbiological analysis.

In both types of experiments, it was clearly shown that water repellents alone do not stop microbial colonisation, while water repellents plus biocides prevent microbial growth.

In addition, it was shown under indoor and outdoor conditions that fungi are able to colonise untreated mortars as well as those treated only with hydrophobic compounds before phototrophic microorganisms.

In a study by Wiktor et al., an accelerated laboratory test was developed to study the biodeteriorative effect of different fungal strains to a cementitious matrix (Wiktor et al. 2009).

The authors used polyethylene boxes $9.5 \times 9.5 \times 9.5$ cm, covered by vermiculite in order to keep humidity inside. A paper sheet was disposed on it to avoid the direct contact between specimens and vermiculite. Two specimens of each matrix of ordinary white Portland cement (water/cement mass ratio 0.55) were placed in each box (Fig. 5.9).

Three fungal strains were selected for the test in order to represent the main kind of fungi involved in biodeterioration in natural environment: *Alternaria alternata (MC342)* to represent a hyphomycete, melanin producer *Exophiala sp. (MC843)* for yeast-like fungi, and *Coniosporium uncinatum (MC557)* as meristematic fungi.

Six specimens were inoculated with 1.5 ml of fungal units suspension (two unweathered, two carbonated, two carbonated and leached), and placed in three

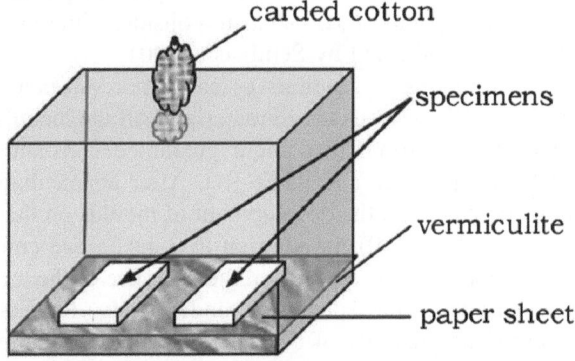

Fig. 5.9 Experimental setup for biodeterioration test. Reprinted from Wiktor et al. (2009), copyright 2009, with permission from Elsevier

carded cotton

specimens

vermiculite

paper sheet

different boxes, control samples were inoculated with 1.5 ml of sterile medium. Boxes were incubated at 26 °C.

In order to monitor fungal growth during the test, the surface of each specimen was observed once a week with stereomicroscope. After 4 weeks of incubation, one specimen of each box was taken and broken into small pieces for Periodic Acid Schiff Staining (PAS) and Scanning Electron Microscopy (SEM). The test permitted to obtain rapid fungal development on cement specimens.

For carbonated specimens, the growth of *Exophiala sp.* was noticed in the third week of incubation. *A. alternata* development started in the first week of incubation, and the growth increases until the fourth week. Hyphae and spore production were observed on the surface of the specimens. Development of *C. uncinatum* was noticed in the second week of incubation and increased until the fourth week.

Stereomicroscopy observations showed that microbial growth was noticed only on the surface of specimens, while PAS staining revealed the real extent of microbial growth on and within the matrix as later confirmed by SEM observations of cross section showing the penetration of hyphae inside the matrix.

Decrease in surface pH increases matrix bioreceptivity considerably. Microbial colonisation was observed on some carbonated specimens and on all carbonated and leached specimens. Carbonation is in fact the most common chemical reaction influencing cement-based materials in natural environmental scenario. That is why accelerated weathering of matrix is generally performed by carbonation (Dubosc 2000; Shirakawa et al. 2003).

Only 3 months of experiments were needed to obtain the first results—mainly related to aesthetical biodeterioration, which is rather shorter than other tests developed to date to study fungal biodeterioration.

D'Orazio et al. analysed the growth rate of three species of mould (*A. versicolor, P. chrysogenum, Stachybotrys chartarum*) on plasters, finishes and paints typically used in heavy weight building envelopes, in order to assess the influence of the substrate chemical composition (in terms of organic fraction of the materials) on the growth rate of moulds (D'Orazio et al. 2009).

The selected materials were two types of plaster (*A, B*), three finishes (*C, D, E*) and two paints (*F, G*) so as to consider different types of support belonging to the classes indicated by Sedlbauer (2001).

Two hundred microliters of each spore suspension were inoculated onto the surface of the various tested materials. All the inoculated caps were then incubated in a climatic room that could guarantee constant environmental conditions of 25 °C temperature and 95 % RH. After an incubation period of 2 weeks in the climatic chamber, the development of moulds on the different materials tested was assessed (Fig. 5.10). In addition to a naked eye comparison, a laser fluorescence microscope was used to reveal the presence of the moulds which had developed on the surface of the specimens under study. Results showed a good correlation between the quantity of organic substances contained in paints, plasters, and finishes and the growth rate of the mould.

Fig. 5.10 Mould formation (*S. chartarum*) visible to the naked eye, over a sample of an interior organic paint, after an incubation period of 2 weeks in the climatic room. Reprinted from D'Orazio et al. (2009), copyright 2009, with permission from Elsevier

References

Adan OCG (1994) On the fungal defacement of interior finishes. Dissertation, Eindhoven University of Technology

Barberousse H, Ruot B, Yéprémian C, Boulon G (2007) An assessment of façade coatings against colonisation by aerial algae and cyanobacteria. Build Environ 42:2555–2561. doi:10.1016/j.buildenv.2006.07.031

Barreira E, Freitas V (2011) Biological defacement of ETICS—a risk assessment methodology. In: XII DBMC International conference on durability of building materials and components, Porto, Portugal

Clarke J, Johnstone C, Kelly N et al (1997b) Development of a simulation tool for mould growth prediction in buildings. In: Proceedings of the Fifth international conference of the international building performance simulation association, Prague, Czech Republic

Clarke J, Johnstone C, Kelly N et al (1999) A technique for the prediction of the conditions leading to mould growth in buildings. Build Environ 34:515–521. doi:10.1016/S0360-1323(98)00023-7

D'Orazio M, Palladini M, Aquilanti L, Clementi F (2009) Experimental evaluation of the growth rate of mould on finishes for indoor housing environments: effects of the 2002/91/EC directive. Build Environ 44:1668–1674. doi:10.1016/j.buildenv.2008.11.004

Delgado J, Freitas V de, Ramos N, Barreira E (2010) Numerical simulation of exterior condensations on façades: the undercooling phenomenon. In: XI International conference on the thermal performance of the exterior envelopes of whole buildings, Clearwater Beach, Florida

Dubosc A (2000) Etude du développement de salissures biologiques sur les parements en béton: mise au point d'essais accélérés de vieillissement. Dissertation, Institut National des Sciences Appliquées de Toulouse

Escadeillas G, Bertron A, Blanc P, Dubosc A (2006) Accelerated testing of biological stain growth on external concrete walls. Part 1: development of the growth tests. Mater Struct 40:1061–1071. doi:10.1617/s11527-006-9205-x

Escadeillas G, Bertron A, Ringot E et al (2009) Accelerated testing of biological stain growth on external concrete walls. Part 2: quantification of growths. Mater Struct 42:937–945. doi:10.1617/s11527-008-9433-3

Hagentoft C-E (2001) Introduction to building physics. Studentlitteratur, Lund (Sweden). ISBN: 91-44-01896-7

Hens H (2003) Mold in dwellings: field studies in a moderate climate. In: Proceedings of the 24th AIVC conference and BETEC conference, ventilation, humidity control and energy, Washington, DC, USA, pp 12–14

Hens HLSC (1999) Fungal defacement in buildings: a performance related approach. HVAC&R Res 5:265–280. doi:10.1080/10789669.1999.10391237

Hukka A, Viitanen HA (1999) A mathematical model of mould growth on wooden material. Wood Sci Technol 33:475–485. doi:10.1007/s002260050131

IEA (1991) IEA-Annex 14. Condensation and energy: source book

IEA (1990) IEA-Annex 14. Condensation and energy: guidelines & practice

Krus M, Fitz C, Holm A, Sedlbauer K (2006) Prevention of algae and mould growth on facades by coatings with lowered long-wave emission. Report, Fraunhofer Institut Bauphysik, Stuttgart

De Muynck W, Ramirez AM, De Belie N, Verstraete W (2009) Evaluation of strategies to prevent algal fouling on white architectural and cellular concrete. Int Biodeterior Biodegradation 63:679–689. doi:10.1016/j.ibiod.2009.04.007

Ojanen T, Viitanen H, Peuhkuri R et al (2010) Mold growth modeling of building structures using sensitivity classes of materials. Thermal performance of the exterior envelopes of whole buildings XI international conference. Clearwater Beach, Florida, pp 1–10

Sedlbauer K (2002) Prediction of mould growth by hygrothermal calculation. J Build Phys 25:321–336. doi:10.1177/0075424202025004093

Sedlbauer K (2001) Prediction of mould fungus formation on the surface of and inside building components. Dissertation, Fraunhofer Institut for Building Physics

Shirakawa M, Beech IB, Tapper R et al (2003) The development of a method to evaluate bioreceptivity of indoor mortar plastering to fungal growth. Int Biodeterior Biodegradation 51:83–92. doi:10.1016/S0964-8305(01)00129-9

Urzì C, De Leo F (2007) Evaluation of the efficiency of water-repellent and biocide compounds against microbial colonization of mortars. Int Biodeterior Biodegradation 60:25–34. doi:10.1016/j.ibiod.2006.11.003

Vereecken E, Roels S (2012) Review of mould prediction models and their influence on mould risk evaluation. Build Environ 51:296–310. doi:10.1016/j.buildenv.2011.11.003

Wiktor V, De Leo F, Urzì C et al (2009) Accelerated laboratory test to study fungal biodeterioration of cementitious matrix. Int Biodeterior Biodegradation 63:1061–1065. doi:10.1016/j.ibiod.2009.09.004

Zheng R, Janssens A, Carmeliet J et al (2004) An evaluation of highly insulated cold zinc roofs in a moderate humid region—part I: hygrothermal performance. Constr Build Mater 18:49–59. doi:10.1016/S0950-0618(03)00025-4

Chapter 6
Nearly Zero Energy Buildings and Proliferation of Microorganisms

Abstract In recent building practice, obligations to legislation on energy saving are carried out mainly by a high thermal resistance and a global airtightness of the envelope, aiming to minimise heat dispersions by conduction and infiltration as much as possible. These measures determine new ways of heat and moisture exchange in the building envelope and are likely to exacerbate the growth of microorganisms. New poorly permeable buildings are in fact more subject to high internal moisture load, in combination with an unsuitable ventilation strategy. Modern exterior insulation finish systems do not have much thermal inertia and are more subject to undercooling phenomena, condensation and a consequent higher biological growth risk. Renovation techniques, such as the replacement of single glazed windows by new very tight double or triple glazed windows or the addition of interior insulation, induce condensation phenomena on the unavoidable thermal bridges (frames, subframes, structure). The NZEB of the future must be able to give a concrete answer to these problems, since, although no changes occur in the thermal performance of the buildings, biological defacement has an enormous aesthetic, health and economic impact, which gathers the disapproval of building's dwellers. This chapter will explore these topics, by describing the major consequences of the 'sealing action' and 'overinsulation' on the proliferation of microorganisms in NZEB.

Keywords Nearly zero energy building · Biological risk · Insulation · Thermal decoupling · Airtightness · Moisture · Undercooling · ETICS

6.1 Why NZEB Could be at Greater Biological Risk?

The need to reduce energy consumption and greenhouse gas emissions, promoted in response to the Kyoto Protocol (in force since 16 February 2005), has led many countries to adopt legislation and strategies to improve the thermal performance of buildings.

E. Di Giuseppe, *Nearly Zero Energy Buildings and Proliferation of Microorganisms*,
SpringerBriefs in Applied Sciences and Technology, DOI: 10.1007/978-3-319-02356-4_6,
© The Author(s) 2013

The European regulatory framework for achieving high energy performances in buildings is based on the Energy Performance of Buildings Directive 2002/91/EC (EPBD) (European Parliament 2002) and its recast (European Parliament 2010). The latter has established several new or strengthened requirements such as the obligation that from 2019 onwards 'all the new buildings occupied and owned by public authorities are nearly zero-energy buildings' (NZEB) and by the end of 2020 'all new buildings are nearly zero- energy buildings'. In the preliminary drafts, the directive was referring to NZEB as 'net zero-energy buildings'. Nevertheless, it seems that the global economic crisis of recent years has prompted lawmakers to scale back targets, considering the fact that 'net zero' is too expensive (Adhikari et al. 2012).

The European Member States 'shall draw up national plans for increasing the number of NZEBs'. Concepts and examples for nearly zero energy already exist in various countries and from different sources. However, the means and techniques to achieve specific national targets show considerable differences across Europe. The definition of NZEB in the EPBD recast offers flexibility, but at the same time leaves uncertainties on the actual ambition level and CO_2-emissions of such buildings.

Since 2008 until the end of 2013, researchers from almost all over the world will have been working together in a joint research program called 'Towards Net Zero Energy Solar Buildings' of International Energy Agency, Solar Heating and Cooling Program Task 40 (IEA) in order to bring the Net ZEB concept to market viability.

The objective of this program is to develop a common understanding, a harmonised international applicable definition framework (Subtask A), design process tools (Subtask B), advanced building design and technology solutions and industry guidelines for Net ZEBs (Subtask C). The scope encompasses new and existing residential and non-residential buildings located within the climatic zones of the participating countries.

In building practice, obligations to legislation on energy saving in buildings are mainly carried out by reducing thermal transmittance and air permeability of the envelope, i.e. by using considerable thickness of insulation and high-performance glazed components. This 'sealing action' aims to minimise air leaks and consequent heat dispersion as much as possible.

In addition to improving energy efficiency, the second major motivation for creating airtight building envelopes is moisture control. Interstitial condensation in building enclosures is primarily the result of forced air exfiltration. Because of the stricter requirements regarding global building airtightness, the installation of air barrier systems and airtight synthetic insulation materials is becoming more and more necessary.

However, the airtightness of the internal environment, if not properly managed with ventilation equipment, may lead to high internal moisture load and therefore surface condensation and mould growth.

The widespread use of synthetic insulation materials, often mounted for insulating the exterior side of the envelope (ETICS) exacerbates the phenomenon of condensation on the surface, with a consequent major biological risk. These building systems do not have much thermal inertia and are therefore more subject to exterior undercooling phenomenon, condensation, and a consequent higher biological growth risk. They are often made of materials that are quite 'vulnerable' such as new organic renders and paints, typical final layers of ETICS, which are particularly susceptible to biological attacks.

In addition, renovation techniques such as the replacement of single glazed windows by new very tight double or triple glazed windows or the addition of interior insulation, induce condensation phenomenon on the unavoidable thermal bridges (frames, subframes, structure) (Fig. 6.1).

All these measures, if on the one hand greatly increase the energy performance of buildings, could on the other hand determine new ways of heat and moisture exchange in the building envelope which are extremely different from those of traditional building envelopes, we have always been accustomed to.

6.1.1 The 'Thermal Decoupling Phenomenon' in NZEB

As already seen in Sect. 3.2, the main moisture loads on façade systems are wind-driven rain and condensation of vapour from outdoor air. The last phenomenon is not only dependent on external climatic conditions, but it is also strictly related to the thermal resistance of the building envelope.

Fig. 6.1 Example of a typical mould formation after the replacement of single glazed windows by new tight windows. View of the new windows installed (on the *left*) and detail (on the *right*)

Exterior condensation on a wall surface occurs when its temperature drops below the dew point of the ambient air. As already described in several papers, the main reason for this temperature drop is the long-wave radiation exchange of the façade with the atmosphere (Künzel and Sedlbauer 2001).

In addition to solar radiation, a building facade is also exposed to long-wave radiation (with a maximum intensity at about 10 μm. The facade itself emits long-wave radiation with an intensity that depends on its emissivity ε and its temperature ϑ (Eq. 6.1):

$$E = 5.67 \cdot 10 - 8 \cdot \varepsilon \cdot (\vartheta + 273, 15)^4 \tag{6.1}$$

where E (W/m^2) is the emitted long-wave energy flux, ε [−] the emissivity, and ϑ [°C] the surface temperature.

Building (non-metallic) surfaces usually have emissivities between ca. 0.8 and 1. Typical long-wave emissions are therefore approximately 300 W/m^2 at 0 °C and 400 W/m^2 at 20 °C.

On the other hand, the facade absorbs part of the long-wave radiation emitted by surrounding objects (terrestrial counterradiation) and by the sky (atmospheric counterradiation). The relative contributions of these two sources depend on the fractional parts they occupy in the field of view of the façade.

A clear sky noticeably emits less radiation than terrestrial objects do. For Central European climate conditions and typical cloud cover, the average radiation intensity emitted by the sky is about 80 % of the intensity emitted by terrestrial objects.

Since the facade of a building is a terrestrial object, a net loss of thermal radiation will occur towards the sky, while the radiation exchange with other terrestrial objects will be roughly balanced. As a result, the long-wave radiation balance of the facade is usually negative, and at night (where no solar radiation can compensate the loss), its surface temperature may drop below the ambient air temperature until convective heat transport from the air towards the facade or any heat flow from indoors counterbalances the radiative loss.

However, in well-insulated buildings, heat flow from the interior of the building is very low. As a consequence, the external surface of highly insulated buildings is generally colder than that of a poorly insulated one (Künzel et al. 2002). If it reaches the dew point of the ambient air, dew formation occurs.

At normal ambient temperature and for the usual night-time relative humidity of 80 % or more, the dew point is only 4 degrees or less below air temperature. Dew formation is therefore common at night.

D'Orazio et al. (2010) described the phenomenon of 'thermal decoupling' between the envelope layers located outside and those inside the insulation in building components with extremely low thermal transmittance. The outer surfaces are most stressed by outside weather conditions, while the inner ones have a quasi-steady behaviour. In Italian temperate climate, in winter, the authors recorded external surface temperatures, which at night usually fall below dew point, keeping the envelope in a situation of perennial surface condensation. As a result, soiling and microbial growth could take place.

6.1.2 Internal Moisture Loads in NZEB

Concerning indoor environments, poorly permeable buildings are more subject to high moisture load, in combination with an unsuitable ventilation strategy.

Internal moisture comes from a wide range of sources: occupant respiration, bathing and showering, cooking, plants. For single-family houses it can range from roughly 4 to 23 l per day (Daquisto et al. 2004). Interior moisture loads in a building can strongly affect indoor humidity levels. Along with surface temperatures, these could cause condensation on interior materials.

Condensation forms on indoor surfaces when their temperature falls below dew point, which varies according to the relative humidity of the air before the drop in temperature and the initial air temperature. For the typical range of relative humidity and air temperature likely to be found in homes, the drop in temperature that will trigger condensation is around 1–5 °C (Galvin 2010). Condensation is therefore a potential problem for any dwelling in which relatively high daytime indoor temperature and relative humidity (caused by indoor human activities) are followed by a temperature drop at night. As moulds need a relative humidity of about 80 % (depending on temperature) (see Sect. 4.2) for a long period of time and algae need higher humidity or even free water for their growth (see Sect. 3.2), periods of wall surface condensation are generally taken as criterion to establish biological development.

The 80 % RH value next to the surface stated by IEA (1990) may seem quite high given that typical indoor RH levels will range from roughly 20 to 60 %. However, relative humidity values in the air adjacent to a surface are strongly affected by air circulation within the space and temperature of the adjacent surface.

The recent trend toward 'tighter' building envelopes could affect building moisture performance by allowing less movement of water vapour via airflow to occur by natural air infiltration. Sealing building envelopes and reducing infiltration can have the effect of increasing indoor moisture levels, especially in climates where interior moisture sources tend to dominate.

To account for the reduction in natural infiltration in NZEB, efforts are being made to increase moisture source control measures, educate homeowners on controlling indoor moisture and to integrate mechanical ventilation in new houses. The exchange of drier outdoor air for more humid indoor air can help to mitigate indoor humidity by essentially diluting the indoor moisture.

ASHRAE's Standard 62.2, approved in July 2003, requires mechanical ventilation in virtually all new houses and reflects the position that ventilation and tight building envelopes are complementary measures (ANSI/ASHRAE 2003).

HVAC systems play a direct role in the control of water vapour through dehumidification, which removes water vapour from the air, and ventilation. Dehumidification usually involves cooling air, thereby reducing its ability to hold as much moisture. Nevertheless, many issues related to HVAC systems performance and design can compromise the effectiveness of this kind of equipment to control indoor humidity. Their design and sizing is very difficult, above all for shoulder seasons in hot, humid climates. When HVAC systems are 'oversized', they are too responsive in meeting sensible cooling loads and as a result are too quick and an inadequate moisture removal takes place.

Supplemental dehumidification is then necessary for new houses once mechanical ventilation requirements start to take hold, because the mechanical introduction of outdoor air will add even more latent load to the indoor environment.

All these kinds of plant equipment for the management of internal moisture load may have very considerable costs, both in the installation phase as well as in during its use.

New buildings can also have condensation problems due to built-in moisture (water contained in some envelope layers is 'sealed' by the insulating layers) and bad construction details (worsening of the phenomenon of thermal bridges).

In recent decades, many initiatives have caught on to save energy also during building renovations. The most common and relatively inexpensive initiative is that of replacing single glazed windows by new, very tight double or triple glazed windows. This renovation technique leads not only in increased air tightness but may also shift the lowest surface temperature from the windows to the outside walls or on to the frames and subframes, inducing condensation phenomena on them (Fig. 6.2). Another renovation technique, which introduces an increased condensation risk, is the addition of interior insulation, since thermal bridges are difficult to avoid.

Fig. 6.2 Example of mould formation in a renovated building whose walls had been recently insulated: the concrete beam of the roof forms a thermal bridge that was not correctly treated

6.2 Proliferation of Microorganisms on NZEB Facades

The formation of microorganisms on the external walls of new or renovated buildings whose envelope has external insulation systems (ETICS) is a problem that has become apparent in recent years. This is mainly due to the phenomena of condensation on the surface, which occurs more frequently on the surfaces of the finishing of insulating panels compared to those of traditional building envelopes (Fig. 6.3). Although no changes occur in the thermal performance of the system, biological defacement has an enormous aesthetic impact, which risks to restrict the full implementation of this technology (Barreira and Freitas 2011).

In general, an ETICS consists of three main components:

1. Insulation. The most common materials are expanded polystyrene foam and mineral wool insulating boards.
2. Reinforcement. In order to reinforce the insulation boards, reinforcing plaster and reinforcing mesh made from glass fibre are applied.
3. Topcoat. Organic renders and paints, plasters, wood, ceramics, glazed bricks and metal can be used for this purpose.

Their application is simple. The insulation boards are fixed to the outside wall with adhesives, anchors or mechanically with rails. After that, they are coated with plaster that has been reinforced with woven glass fibre mesh. The final layer consists of the finishing plaster or the desired surface material.

Lengsfeld and Krus observed the external surface temperature course of an old building (U-value $= 1.10$ W/m^{2}K) and of a well-insulated façade (U-value $= 0.35$ W/m^{2}K) in Autumn (Lengsfeld and Krus 2004). The surface

Fig. 6.3 ETICS defacement due to biological growth (buildings located in Porto, Portugal). Reprinted from Barreira and de Freitas (2013), copyright 2013, with permission from Elsevier

temperature of the old building was higher than the dew-point temperature during the whole night: the higher thermal transmittance leads to a higher outgoing heat flux, which increases the surface temperature.

The temperature of the well-insulated façade decreased under the dew-point temperature from 3 to 8 °C in the morning. The phenomenon is due to the 'thermal decoupling' that occurs in the envelope: the layers of the wall inside the insulation are greatly affected by internal environmental conditions, while, as the heat flows through the wall are more limited, the layers outside the insulation are almost only subject to the external conditions of temperature and relative humidity. The result was a higher amount of condensation on the surface of the well-insulated façade and a consequent higher biological growth risk.

Another consequence of adding external insulation layers on facades is a slower drying process of liquid water accumulated on the surface due to surface condensation and wind-driven rain, because of the diminished exterior surface temperature (Kuenzel and Sedlbauer 2001). The drying process allows the evaporation of the liquid water accumulated on the surface. Evaporation from the wet surface occurs whenever the saturation pressure at the surface is greater than the vapour pressure of the ambient air. If the drying process is not sufficiently fast, the surface moisture content remains high for long periods and increases the risk of microbiological growth (Krus et al. 2006).

Künzel et al. carried out periodic inspections of buildings with External Thermal Insulation Composite Systems (ETICS) in Central Europe, from 1975 to 2004 and classified the state of the façades into three assessment groups: Group 1—Virtually without Defects, Group 2—Minor Defects and Group 3—Major Defects (Künzel et al. 2006). The development of algae was not assessed as technical defect but as a 'visually adverse effect.' The authors detected a great susceptibility of ETICS to microbial growth due to rain or night-time surface condensation. In some cases, the formation of algae was visible, while in other cases, the cleaning effect due to rain prevailed.

Analysis of the hygrothermal behaviour of an ETICS based on mineral wool at three different European locations performed by the simulation tool for the transient heat and moisture transfer WUFI[®][1] (Zirkelbach et al. 2005) showed that strongly varying moisture strains can occur in the mineral wool insulating layer of an ETICS. The highest values of relative humidity and temperature can be observed at the interface between the mineral wool and the exterior render—independently of the three examined climate locations.

The heat capacity of a finishing material has a high influence on temperature variations of the envelope. If the air temperature decreases rapidly (at nights), the surface temperature on the outer layer of a construction with high heat capacity will be warmer than the air temperature. If the outer layer of the wall has low heat

---

[1] WUFI[®] (Wärme und Feuchte instationär) is a software family, which allows realistic calculation of the transient coupled one- and two-dimensional heat and moisture transport in multi-layer building components exposed to natural weather.

capacity, the temperature of this surface will be closer to the air temperature. This means that the relative humidity on the surface of the outer layer with low heat capacity will be higher and increases the risk for microbial growth. Calculations of surface temperature on different walls show that the temperature on façades with low heat capacity is about 0.5 K lower at night compared with walls with high heat capacity (Johansson et al. 2005). Modern exterior insulation finish systems do not have much thermal inertia and are then more subject to considerable amounts of exterior condensation (Künzel and Sedlbauer 2001).

Johansson has carried out several investigations on experimental and real buildings to evaluate the growth of microorganisms on the façade (Johansson et al. 2010; Johansson 2006).

Recently, he monitored surface temperature and surface relative humidity over a 20 month period on a test house with different façade elements: one "heavy" and one "light" (Johansson et al. 2010). The light system consisted of (from the outside) a thin rendering on polystyrene insulation and the interior insulated wood frame. The heavy wall had a thin rendering on a brick wall placed outside the polystyrene insulation and the insulated wood frame. Both construction types had the same type of organic rendering system and the same total thickness of the thermal insulation layer (Table 6.1).

In order to investigate the influence of façade colour, half of each façade element had a final rendering layer with red pigments, whereas the other half had a white-pigmented final rendering.

Results showed that thin renderings on thermal insulation have significantly higher surface humidity compared to façade constructions with higher thermal inertia, and therefore have a higher potential for mould growth. The colour is the most important factor for surface humidity levels on south-facing façades (in the northern hemisphere) as darker surfaces absorb more solar radiation and therefore have a higher average temperature. On a north-facing façade, the heat storage capability of the façade and its effect on the surface temperature is most important.

Using ETICS for new or renovated building also requires skill workers for installation in order to prevent bad construction details inducing moisture damage.

By observing northeast-facing walls of experimental buildings with ETICS (10 cm exterior insulation, $U = 0.3$ W/m^2 K), Künzel and Sedlbauer noticed that the typical spotted pattern caused by microbial growth presented some clear spots of about 5 cm diameter, where algal growth was not visible (Kuenzel and

Table **6.1** Stratigraphies of the constructions of the light (*left*) and heavy (*right*) façades studied by Johansson et al. (2010)	Light wall	Heavy wall
	3 mm organic rendering	3 mm organic rendering
		Bricks 120 mm
	50 mm extruded polystyrene	50 mm extruded polystyrene
	9 mm gypsum board	9 mm gypsum board
	145 mm mineral wool	145 mm mineral wool
	0.2 mm vapour barrier	0.2 mm vapour barrier
	9 mm gypsum board	9 mm gypsum board

Sedlbauer 2001). These bright spots resulted from the thermal bridging of the fixing anchors beneath the stucco. Their temperature exceeded the rest of the façade by about 1.5 K. In fact, the anchors cause a higher heat flux on the surface than the insulation, thus preventing the moisture supply for biological growth.

Sedlbauer and Krus detected a noticeable biological growth on the external facades of a housing project after a short time by the end of its construction, especially in the area around the lintels (Sedlbauer and Krus 2002). It turned out that the insulation slabs of expanded polystyrene had not been applied without joints, but that a continuous gap of approximately 3 mm existed between the insulation sheets. The patterns of mould infestation were approximately located in the area of the joint crosses where four insulation slabs met.

Johansson investigated several buildings with discoloured façades, mostly in the southernmost part of Sweden, also noticing organised patterns of small round areas where no growth occurred on otherwise quite-fouled façades (Johansson 2006). Further investigations showed that the 'white dots' appeared at each fastener placed under the rendering to fasten the thermal insulation to the structural wall behind. The fasteners were heat bridges, with higher outer surface temperature and therefore with lower surface relative humidity at which no microbial growth occurred. Results from computer simulations showed that the fasteners in cases with thin layers of thermal insulation and rendering give a 1–2 K higher night-time temperatures than the rest of the façade, which made a difference in relative humidity of 11 %. This small difference in temperature led to differences in moisture state that were high enough to almost prevent microbial growth at the fasteners.

In addition to the problem of the algae formation on ETICS, the mould growth in the space between the insulation panels and the outer wall as a result of high humidity level can also appear as a risk (Antonyová et al. 2013). In building renovations, the thermal insulation of envelope is often implemented through partial bonding of EPS sheets. If the external surfaces is very irregular and uneven, or insulation procedure is not conducted by professional implementation and the building is not properly heated during the winter, the insulation plates could be about 1 cm or more distant from them. In this way, an empty space is created between the wall and the panels, where humid air can be trapped. The gap could then become a suitable environment for growth of spores and moulds.

6.3 Mould Growth Inside NZEB

As already seen in previous sections, although it may be reasonable to consider the 80 % RH of a material as threshold level for growth in buildings, the real threshold varies that depend on several factors. They are related to the environmental conditions (temperature, relative humidity), the type of substrate (porosity, adsorption curve, capacity of maintenance of the conditions) and other secondary factors (the ageing of the materials themselves, the presence of species of different fungal, cleaning and maintenance of the substrate materials). Consequently, the relationships that exist between construction technologies and mould growth are still to be further studied in depth, also taking into account the constructive innovations related to NZEB.

According to Hens, nine parameters define the likelihood for to develop: (1) climate, (2) inside temperature, (3) vapour release, (4) ventilation, (5) layout, (6) envelope thermal performance, (7) moisture buffer capacity inside, (8) presence of preferential condensation surfaces and (9) type of finish (Hens 2003). Exterior climate acts as boundary condition while inside temperature, vapour release and ventilation belong to living habits. The five others parameters are design and construction related.

Hens analysed 35 cases in real buildings from 1972 to 2002 to evaluate the influence of the nine parameters (Hens 2003). Lack of ventilation, large surfaces of exterior walling, low inside temperatures and poor envelope thermal performance ranked high. A low inside temperature is typically the result of a rebound effect. The fact that the dwelling has such poor thermal performance is because people accept lower comfort to keep heating payable. Ventilation in turn is to a large extent building related. In most cases, new residential buildings are too airtight for adventitious ventilation; ventilation system is not designed or, if present, not used. In insulated dwellings, thermal bridging is the main cause.

The latest constructive trends conceive buildings more and more as 'airtight boxes', either due to a reduced permeability of the envelopes and windows for the purposes of a reduction of the heat loss, both due to the diffusion of plasters, materials for interior coverings and painting with poor hygroscopic properties, but fast and economic application. Such measures in combination with an unsuitable ventilation strategy, risk to bring about higher moisture load and to cause the phenomenon of microorganism growth indoors.

In Europe many energy buildings certifications require very high airtightness standards. For example the 'Passive House'-standard[2] and the 'Minergy-P'-standard[3] explicitly require a building airtightness of 0.6 air changes per hour (ACH) at 50 Pa (Langmans et al. 2012).

In timber frame constructions, airtightness is traditionally created by an interior 'air barrier system'. In cold and moderate climates, such as North-West European areas, this air barrier function is often combined with that of the 'vapour retarder'. In order to protect the insulation layer from infiltration of outside cold air, a 'wind barrier' is often provided outside the insulation (Uvslokk 1996). In addition, this exterior layer serves as drainage plane to prevent water infiltration into the structure.

As the installation of air barriers on the inner side of the envelope is difficult (due to the presence of structural nodes, plant cables and pipes), the recent trend is

---

[2] Passivhaus Standard is a German voluntary standard for the energy efficiency in buildings.

[3] Minergy-P is a Swiss registered quality label for new and refurbished low-energy-consumption buildings.

to move the interior air barrier to the outside of the building envelope where fewer joints are present. Recent studies show the advantage of improving the airtightness of the wind barrier, so it will serve as an exterior air barrier system (Langmans et al. 2012). In this case, the exterior sheathing acts as a wind and air barrier and the inner sheathing acts only as a vapour barrier/retarder. Nevertheless, given the non-continuous interiors heating and other transport mechanisms such as internal natural convection might be introduced: inside air with a higher absolute humidity can enter the insulation causing interstitial condensation or growth against the airtight exterior sheathing.

Langmans et al. (2012) studied highly insulated timber frame walls with an exterior air barrier, enclosed between two climate chambers (hot box/cold box) to simulate winter conditions in a temperate climate. The walls differed from each other by the physical properties of the applied exterior air barrier: the airtightness, moisture buffer capacity, vapour permeability and thermal resistance of the air barriers varied.

Results showed an increased moisture flow at the upper part of the walls, highly depending on air permeability and the accuracy of the installation of the insulation layer. The phenomenon is as much damaging as long lasting: this convection loop provides a constant moisture supply towards the upper cold side of the structure, thereby improving conditions for mould growth.

Their study showed good results by using exterior sheathing materials with high thermal resistance and moisture buffer capacity. Thermal resistance increased the temperature and thus reduced the relative humidity on the cold side of the insulation. Therefore, the study showed that moisture buffer capacity could delay condensation conditions.

Mlakar and Štrancar (2013) compared three different lightweight test passive houses (A, B and C) and the associated temperature and relative humidity profiles. The humidity experiments were performed with humidifiers to humidify all the houses for 1 h per day in order to emulate rooms in real houses where the amount of water vapour increases for the activities of occupants (cooking, showering, etc.).

Results showed that after 10 days of experiments, the houses responded differently to the humidity load, based on their construction. The relative humidity in houses A and B (house A had cellulose insulation and house B had wood-fiber board) increased only by around 5 % in 10 days, while in house C (with mineral wool and a vapour barrier) the increase was around 30 %. The vapour barrier installed prevented any vapour exchange with the outdoor environment, and therefore all the vapour produced remained in the house. In addition, mineral wool is not hygroscopic and cannot balance the internal relative humidity variations. In a real house, this would mean an excess of water vapour to be removed with ventilation, which would increase energy consumption.

In the other two houses, there was no vapour barrier. This meant that their walls were still permeable to water vapour. Moreover, their insulation layers were composed of good hygroscopic materials that could buffer the indoor humidity oscillations. All these factors kept the internal relative humidity in houses A and B at comfort level. In reality, this would mean more comfortable relative humidity levels and less possibility of humidity-related fungus growth, condensation and deterioration of the building materials.

Attic constructions can be frequently found in Northern Europe and North America. They usually have insulation on a horizontal ceiling with an accessible space above. The low price and reasonably low maintenance requirement are some of the reasons for the selection of cold pitched roofs in house designs.

According to the recent needs for energy saving in buildings, many residential attics have been insulated by adding extra insulation in the ceiling. Due to these high insulation levels, attics have become colder and more susceptible for moisture (Nik and Sasic Kalagasidis 2013). The main sources of moisture in cold attics are the infiltration of moist indoor air up to the attic through the attic floor and construction dampness (built-in moisture). These kind of problems in unheated zones of buildings such as cold attics have increased remarkably over the last decade. The latest filed survey of the Swedish National Board for Housing and Planning, which enclosed 1,400 residential buildings sampled all over the country, showed that the mould growth was present in 21 % of attics (Nik and Sasic Kalagasidis 2013). The appearance of mould and durability concerns of attics have been also increased with respect to future climate changes (Kjellström et al. 2007). Climate predictions point to warmer and more humid climate, which may increase the number of moisture-related problems in attics.

Nik and Sasic Kalagasidis investigated the hygrothermal performance, mould index and its growth rate, of an attic with four different roof constructions (no insulation, insulated, with gable side ventilation and with mechanical ventilation) in different climate change scenarios in Sweden (Nik and Sasic Kalagasidis 2013). According to the Matlab[4] simulation results, temperature and humidity levels will increase in the attics if 'climate change' happens, and that may increase the risk of growth in the attics. While assessing the future performance of attic constructions, it was seen that the absolute safe case for preventing growth would be to use mechanical ventilation in the attic. However, this method would increase the energy consumption of the building. Attics with insulated roofs can be a solution to the problem with the current climate situation in Sweden. However, they do not show a promising performance if the 'climate change' happens.

Since the average annual rate for the construction of new dwellings in EU-15 member states is about 1 % of the existing stock (Hartless 2003), an improved energy efficiency of the existing buildings is expected to play a key role in meeting the EU commitment to the Kyoto Protocol.

Substantial energy savings of up to 90 % could be obtained when insulation is applied to the exterior building façade (Morelli et al. 2012). This is readily feasible when buildings are built with cladding that could be easily removed or buildings having facades not worthy of preservation. However, in instances where historical and cultural preservation of the exterior of the building facade is of interest,

---

[4] MATLAB (matrix laboratory) is a numerical computing environment and programming language developed by MathWorks.

retrofitting the exterior is evidently not an option, and the only alternative is installing insulation on the interior side of the external wall.

In the energy retrofitting experience of a typical old Danish multi-family building to a 'nearly-zero' energy building carried out by Morelli et al., much attention was paid to a proper installation of the interior insulation on the wall assembly (Morelli et al. 2012). In fact, an incorrect mounting can be a source for mould growth between the insulation and the wall substrate, in particular on brick masonry given their large capacity for moisture uptake and retention, and for moisture problems in the wooden support beams of the floor. Therefore, before applying the insulation, the walls were first cleaned so no organic material was present on the wall.

Authors calculated that in that case study, insulation placed on the interior portion of walls reduced energy consumption by 20 %. However, evident drawbacks were: the loss in living space, a change in the moisture balance of the walls, risk of mould growth on the wall behind the insulation and consequent maintenance costs.

References

Adhikari RS, Aste N, Del Pero C, Manfren M (2012) Preface. Energy Procedia 14:1. doi:10.1016/j.egypro.2011.12.887

ANSI/ASHRAE (2003) STANDARD 62.2-2003 Ventilation and acceptable indoor air quality in low-rise residential buildings

Antonyová A, Korjenic A, Antony P et al (2013) Hygrothermal properties of building envelopes: reliability of the effectiveness of energy saving. Energy Build 57:187–192. doi:10.1016/j.enbuild.2012.11.013

Barreira E, de Freitas VP (2013) Experimental study of the hygrothermal behaviour of External Thermal Insulation Composite Systems (ETICS). Build Environ 63:31–39

Barreira E, Freitas V (2011) Biological defacement of ETICS—a risk assessment methodology. XII DBMC international conference on durability of building materials and components, Porto, Portugal

D'Orazio M, Di Perna C, Di Giuseppe E (2010) The effects of roof covering on the thermal performance of highly insulated roofs in Mediterranean climates. Energy Build 42:1619–1627. doi:10.1016/j.enbuild.2010.04.004

Daquisto D, Crandell J, Lyons J (2004) Building moisture and durability past, present and future work. Report, U.S. Department of Housing and Urban Development, Washington, D.C.

European Parliament (2002) Directive 2002/91/CE of the European Parliament and of the Council of 6 December 2002 on the energy performance of buildings

European Parliament (2010) Directive 2010/31/EU of the European Parliament and of the Council of 19 May 2010 on the energy performance of buildings (recast)

Galvin R (2010) Solving mould and condensation problems: a dehumidifier trial in a suburban house in Britain. Energy Build 42:2118–2123. doi:10.1016/j.enbuild.2010.07.001

Hartless R (2003) Application of energy performance regulations to existing buildings. Final Report of the Task B4, ENPER TEBUC, SAVE 4.1031/C/00-018/2000, Watford, UK

Hens H (2003) Mold in dwellings: field studies in a moderate climate. In: Proceedings of the 24th AIVC conference and BETEC conference, ventilation, humidity control and energy, Washington, DC, USA, pp 12–14

IEA (1990) IEA-Annex 14. Condensation and energy: guidelines & practice

IEA, Solar Heating and Cooling Program (SHC) Task 40—towards net zero energy solar buildings. http://task40.iea-shc.org/. Accessed 9 May 2013

Johansson S (2006) Biological growth on mineral façades. Dissertation, Lund University, Sweden

Johansson S, Wadsö L, Sandin K (2005) Microbial growth on buildings facades with thin rendering on thermal insulation. 7th Nordic building physics symposium, Reykjavik, Iceland

Johansson S, Wadsö L, Sandin K (2010) Estimation of mould growth levels on rendered façades based on surface relative humidity and surface temperature measurements. Build Environ 45:1153–1160. doi:10.1016/j.buildenv.2009.10.022

Kjellström E, Bärring L, Jacob D et al (2007) Modelling daily temperature extremes: recent climate and future changes over Europe. Clim Change 81:249–265. doi:10.1007/s10584-006-9220-5

Krus M, Rosler D, Sedlbauer K (2006) New model for the hygrothermal calculation of condensate on the external building surface. In: Proceedings of the third international building physics conference—research in building physics and building engineering, Montreal (Canada), pp 329–333

Künzel H, Sedlbauer K (2001) Biological growth on stucco. Performance of exterior envelopes of whole buildings, VIII International conference, Clearwater Beach, FL, pp 1–5

Künzel HM, Schmidt T, Holm A (2002) Exterior surface temperature of different wall constructions comparison of numerical simulation and experiment. In: Proceedings of 11th symposium for building physics. Dresden, Germany, pp 441–449

Künzel H, Künzel HM, Sedlbauer K (2006) Long-term performance of external thermal insulation systems (ETICS). ACTA Architectura 5:11–24

Langmans J, Klein R, Roels S (2012) Hygrothermal risks of using exterior air barrier systems for highly insulated light weight walls: a laboratory investigation. Build Environ 56:192–202. doi:10.1016/j.buildenv.2012.03.007

Lengsfeld K, Krus M (2004) Microorganism on façades-reasons, consequences and measures. Report, Fraunhofer Institute for Building Physics, Holzkirchen, Germany

Mlakar J, Štrancar J (2013) Temperature and humidity profiles in passive-house building blocks. Build Environ 60:185–193. doi:10.1016/j.buildenv.2012.11.018

Morelli M, Rønby L, Mikkelsen SE et al (2012) Energy retrofitting of a typical old Danish multi-family building to a "nearly-zero" energy building based on experiences from a test apartment. Energy Build 54:395–406. doi:10.1016/j.enbuild.2012.07.046

Nik VM, Sasic Kalagasidis A (2013) Impact study of the climate change on the energy performance of the building stock in Stockholm considering four climate uncertainties. Build Environ 60:291–304. doi:10.1016/j.buildenv.2012.11.005

Sedlbauer K, Krus M (2002) Mold Growth on ETICS (EIFS) as a Result of "Bad Workmanship"? J Build Phys 26:117–121. doi:10.1177/0075424202026002782

Uvslokk S (1996) The importance of wind barriers for insulated timber frame constructions. J Build Phys 20:40–62. doi:10.1177/109719639602000105

Zirkelbach D, Holm A, Künzel H (2005) Influence of temperature and relative humidity on the durability of mineral wool in ETICS. X DBMC International conference on durability of building materials and components, Lyone, France

Chapter 7
Remedial Actions and Future Trends

Abstract The aesthetic quality and durability of external building envelope could be seriously impaired by the development of microorganisms, which will colonise building materials whenever a suitable combination of dampness, light and "bioreceptivity" of the substrate occurs. The control of biodeterioration in buildings includes measures useful to eliminate the presence of microorganisms and, when possible, to delay their recurrence. The difficulty lies in applying methods that are effective against biodeteriogens but that do not have interaction with the materials of the substrates. This chapter outlines some of the consolidated or innovative approaches which aim to give a concrete answer to the biological problem in buildings, acting both on the microorganisms already disseminated and on the main causes of development. Several methods may be used, in function of the type of organism present, the materials of the substrate and its state of preservation, the construction methods of the building and the freedom and economy of the intervention. Among the traditional methods, mechanical, chemical and physical strategies for the removal of biodeteriogens have been mentioned, while a more detailed study will be done on the use of biocides and water repellents that directly act on the material to prevent it from becoming fertile ground for microorganism development. Among the innovative methods, the use of engineered nanoparticles as additives to envelope finishing materials is catching on. Strategies that include a set of practical design, construction and use of buildings, which allow acting on the environmental conditions that favour the proliferation of microorganisms will be finally reported as sustainable actions.

Keywords Bioreceptivity · Biocide · Hydrophobic compound · Nanoparticles · Titanium dioxide · Phase change materials · Moisture buffering · IR paint

E. Di Giuseppe, *Nearly Zero Energy Buildings and Proliferation of Microorganisms*, SpringerBriefs in Applied Sciences and Technology, DOI: 10.1007/978-3-319-02356-4_7, © The Author(s) 2013

7.1 Traditional Methods

With regard to the principles and nature of the means employed, traditional direct methods can be classified as mechanical, physical and chemical.

7.1.1 Mechanical Methods

Traditional mechanical methods involve the physical removal of biodeteriogens from all types of substrate, either by hand or with tools such as scalpels, spatulas, scrapers, air abrasive or vacuum cleaners. Although frequently used in the past, these methods do not produce lasting results. Moreover, mechanical methods can damage the substrate, even if they have the advantage of not adding on it any substance that might cause further deterioration. In some cases, a preliminary biocide treatment is surely advantageous to facilitate the removal of the biological growth (Tiano 2002).

Furthermore, after surface cleaning interventions some chemical substances (protective coatings or consolidants) can be applied to the object in order to increase its water repellence and cohesion. These products may prevent or retard the recolonisation of the substrate, decreasing its porosity, roughness and water content, but in critical environmental conditions, they could even favour the development of micro flora.

7.1.2 Chemical Methods

Biocides are chemicals used for killing undesirable biological growth because they have a specific toxicity for the species to be eliminated. They are classified in different ways depending on their chemical nature, on the target pest species, or on their mode of action. In particular, we can distinguish the biocides applied before other treatments to eliminate microorganisms already present, and those that should have a preventive effect that slow down the colonisation of surfaces.

Biocides can be added to the liquid paint both to protect it and to reduce fouling of the dry film (Gaylarde et al. 2011). These products must have a certain degree of water solubility in order to leach from the surface and act on the microbial cells. However, they can easily be washed out by rainwater running down the façade, losing their action and contributing to pollution in the soil or water (Bester and Lamani 2010). In Europe, the European Biocidal Products Directive (BPD) (European Parliament 1998) requires an environmental risk assessment of biocidal products in the market, which is based on the comparison of expected concentrations of active ingredients in the environment and ecotoxicity data of these substances.

Special care must be taken in the choice of suitable products, which vary depending on biodeteriogens present as well as the correct dosage and application procedure.

The basic requirements of a biocidal product are:

1. High efficacy against biodeterioration;
2. The absence of interference with the constituent materials;
3. Low toxicity to human health;
4. Low risk of environmental pollution.

A biocide becomes more effective when the dose required is reduced, its spectrum of action becomes wider and the persistence of action is much longer than before. Factors that normally increase the effectiveness of a treatment are: the concentrations used, the duration of the application, the stability of the product and the ambient temperature. Factors that normally reduce it are: the extent of colonisation, the presence of organic material, the presence of cracks in the substrate and the conditions of wind or rain during the treatment. Factors affecting in different ways depending on the active principle are: the type of substrate, its water content, the type of solvent, the pH of the solutions, the light intensity and the rains subsequent to treatment.

As regards the interference of biocidal products with the substrate, it can arise through macroscopic effects, such as stains, yellowing or bleaching, increased brightness or opaqueness, or microscopic effects, such as increased porosity, crystalline microerosion, and modification of the mechanical, chemical and physical properties. Nowadays, different chemical treatments are available on the market for the prevention of biological stains. Among the principal classes of biocides, we can find:

(1) *isothiazoline-3-one* derivatives, i.e. *methyl-isothiazolinone (MI), chloromethyl-isothiazolinone (CMI), benzo-isothiazolinone (BIT), N-octyl-isothiazolinone (OIT)* and *dichloro-N-octylisothiazolinone (DCOIT)*;
(2) *phenylurea* derivates, i.e. *diuronand isoproturon;*
(3) *benzimidazole (carbamate)* derivatives, i.e. *carbendazim, iodocarb;*
(4) *triazines*, i.e., *terbutryn* and *cybutryn (Irgarol$^{®}$ 1051)*;
(5) *thiocyanatomethylthiobenzothiazole (TCMBT).*

Isothiazolinone-based biocides are among the most widely used biocides for paint protection film (Gaylarde et al. 2011). At the moment, *carbendazim* is one of the key biocides in dry film protection, but increasing limitations are being put on the use of this biocide and other products, by European Union directives (European Parliament 2008).

In addition to the pesticide, other ingredients, such as carriers or additives to improve the effectiveness of the product or to facilitate its application, are present in the chemical formulation. These blends include combinations of fungicides and algaecides, such *carbendazim* plus *diuron*, and combinations of fungicides to correct deficiencies in the fungicidal spectrum of each of them, such as *carbendazim* plus *OIT*.

It has been proposed that the application of hydrophobic compounds and biocides together is more effective against microbial colonisation of surfaces. The use of water repellents aims at decreasing the bioreceptivity of the material, while the use of biocides aims at decreasing its biological activity (Urzì and De Leo 2007).

Urzì and De Leo (2007) collected experimental data both in laboratory conditions and outdoors on artificially infected mortars and reference mortars, so as to evaluate the effects of three hydrophobic compounds (*rhodorsil RC80, hydrophase superfici* and *hydrophase malte*), applied alone or in combination with the biocide *algophase* and with the new water miscible formulation (*algophasePH025/d*).

Results demonstrated that untreated mortars possessed a high primary bioreceptivity and were a suitable surface for microbial colonisation by heterotrophic and phototrophic microorganisms. In particular, the two types of mortars used (with a slightly different porosity) showed a different rate of deterioration processes and higher colonisation. In both types of experiments, it was clearly shown that water repellents alone do not prevent biofilm growth on the surface. On the contrary, it seems evident that the combination of water-repellent compounds and biocides applied in a single step creates unfavourable conditions for microbial growth for up to 15 months of incubation under laboratory conditions. No significant differences were observed among the compounds tested in their efficacy in preventing colonisation.

Fungi were able to colonise untreated mortars as well as those treated only with hydrophobic compounds before phototrophic microorganisms. This fact can be explained by capacity of fungi to grow at lower water availability than algae, cyanobacteria and bacteria. In addition, ubiquitous fungi such as *Alternaria*, *Cladosporium* and *Ulocladium* possess a very high metabolic versatility that together with their morphological characteristics allows them to survive in dry and oligotrophic conditions (Urzì and De Leo 2007).

The application of a water repellent and a biocide in the same solution seems to be very effective as also suggested by other studies (Balzarotti-Kammlein 1999; Urzì et al. 2000). Inhibition of microbial growth by biocide treatment can persist for a period of 5–10 years except under continuous conditions of rising dampness. Under these conditions massive phototrophic-based microbial community were observed.

Unfortunately, due to the increase in costs it seems that treatments using water repellents and biocides in combination are rarely suggested. Furthermore, due to comparable efficiency of the water miscible products and due to the directive of European Community 2004/42/CE (European Parliament 2004), which limits the use of volatile organic compounds (VOC) as solvents, the use of compounds with water as solvent should be preferred, where their applications are harmless to the substrate (Urzì and De Leo 2007).

Shirakawa et al. (2002) studied the fungal colonisation and the effectiveness of biocide incorporation on paint films, on two buildings painted with an acrylic paint, with and without an experimental biocide formulation containing acarbamate (*carbendazin*), *N-octyl-2H-isothiazolin-3-one* and *N-(3,4-dichloro-phenyl)N,N-dimethyl urea* (total biocide concentration 0.25 % w/w). The fungal population on biocide-containing surfaces was significantly lower than on non-biocide-containing paint after 13 weeks and continued so to 42 weeks after painting.

Shirakawa et al. (2010) also studied painted concrete panels exposed to various environments in Brazil for 4 years and demonstrated that climate was more important than dry film biocide (*carbenzadim* and *isothiazolinone*). Warkentin et al. (2007), in Germany used accelerated biodeterioration tests on a complete façade showing that the hydrophobicity of the surface was more important than biocide content.

For organic renders, as the ETICS external layers, even more care should be taken when biocide treatments are required, and these must be applied only as the last resource. The application of biocides is foreseen in very special cases, especially when the material is heavily attacked.

The major biocide groups for these types of coatings are: *phenylureas* (*diuron* and *isoproturon*) acting as algaecides, *triazines* (*terbutryn* and *Irgarol1051*) acting as algaecides, *carbamates* (IPBC) acting as fungicide and *isothiazolinones* (*OIT* and *DCOIT*) for controlling microbes in general (Wangler et al. 2012).

The paint industry is now working on different strategies to create a new generation of paint formulations, which are fulfilling the BPD and are able to prevent the paints from microbial colonisation for many years (Kaiser et al. 2013).

One approach is the production of paint formulations with degradable biocides, which can be released only in small quantities. They will not accumulate in the environment and will be sufficient for a long lasting protection of the paints against microbial colonisation. Other strategies look into the possibility to replace the degradable biocides based on organic compounds by nanomaterials, as we shall see later.

7.1.3 Physical Methods

Physical methods may be a viable alternative to biocidal products where these have proven to be ineffective or have manifested interference with the substrate. Through physical methods, one can also avoid the use of products which may leave toxic residues and do not incur the possible effects of environmental pollution. Some methods used experimentally are ultraviolet rays (UV) and gamma rays.

Ultraviolet rays have especially been used against bacteria, algae and fungi in the treatment of renders and plasters. The part of the UV spectrum with germicidal activity is between 300 and 200 μm, with a maximum of activity between 275 and 230 μm. Microorganisms vary in sensitivity depending on their growth phase and the nature of the substrate on which they are found. Disadvantages of this method are that UV radiation is effective at low RH values (less than 50–60 %); it has poor penetration power and can modify some materials and colours of surfaces.

Currently, various manufacturers are marketing germicidal UV lamps for controlling contamination, including fungal contamination in indoor environments, as well as Air Handling Units (AHU's) and ducts (Haleem Khan and Mohan Kar-uppayil 2012).

Gamma rays are a form of electromagnetic radiation. They are especially used on organic materials. Moulds are less sensitive to radiation than insects, and different strains show different levels of sensitivity; generally most fungi are killed by a total dose of 10 kiloGrays (kGy) (Tiano 2002).

Among the physical methods recently used for the treatment of biodeteriogens, we distinguish those based on laser and those based on the absorption of microwave energy.

Laser allows a quick and focused treatment, but it is by its own nature very superficial.

By contrast, microwaves allow—in principle—treatment of regions which are more extended and of greater depth. The disinfestation by means of microwave heating of living organisms has been used in various fields with different degrees of success, depending on the specific thermal characteristics of the organism treated and the characteristics of the substrate that hosts it.

Microwaves have a debioting action. They act on the existing biodeteriogens and act only on moisture reduction and on the substrate. They do not heat the air but act directly on and in the material. In particular, the water has a very high absorption at microwave frequencies, allowing to heat even at a discrete depth an object rich in water without the surrounding material to warm appreciably (if not for thermal conduction).

The apparatus that transfers microwave power (applicator) must be chosen according to the treatment, with the aim to induce "selective" heating which causes a thermal increase in the infested area with a minimum stress of the surrounding material.

7.2 A Promising Prospect: Innovative Engineered Nanoparticles

One of the most promising strategies as an alternative to the use of conventional biocides is the development of innovative materials that use nanoparticles for conferring new functions on traditional materials. "Nanoparticles" are engineered particles with at least one dimension in the range of 1–100 nm (ISO/TS 27687 2008). They improve the paint properties, such as water repellence, scratch resistance, durability and antimicrobial properties (Kaiser et al. 2013). The advantage, in contrast to water-soluble biocides, is that they can be fixed more effectively in the coating matrix, and the threat of leaching out into the surrounding environment is lower, thereby resulting in lower ecological risks.

Nanoparticles in façade coatings can be used as hardener (silica dioxide), UV-light absorber (titanium dioxide or zinc oxide) or biocide (nanometals) in addition to the conventionally applied pigments.

Moreover, the increasing usage of engineered nanoparticles enhances the probability for people to get into contact with them. Eventual adverse health effects are still under study (Som et al. 2011).

Among nanomaterials, there are TiO_2 (titanium dioxide)-based coatings, which provide improved or new properties to paints in order to optimise their rheological or mechanical properties or to give them self-cleaning properties through dirt repellent, photocatalytic and superhydrophylic properties.

When stimulated by UV light, TiO_2 can inhibit some microorganism biofouling, thanks to their photocatalytic properties (Fonseca et al. 2010) (Fig. 7.1). TiO_2 is highly reactive not only because it oxidases organic substances but also because it inactivates bacteria and viruses. However, its effect on algae and, to some extent, on fungi is more doubtful (Gaylarde et al. 2011).

TiO_2 exists as three mineral structures, but only two forms are normally used for architectural purposes. "Rutile" is used as a white pigment in architectural paints because of its high refractive index, resulting in high opacity and whiteness; while "anatase" is mainly used for the protection of building materials. Due to its high redox potential and band gap, the anatase variety of titanium dioxide, in the form of nanocrystalline powder, is one of the most widely used semiconductors for photocatalysis processes. The fact that this compound is non-toxic, very photoactive, photostable and produces colourless films when applied to materials, is a benefit. The incorporation of photocatalysts to construction materials (cement, mortars, exterior tiles, glass) also confers self-cleaning properties to them: the

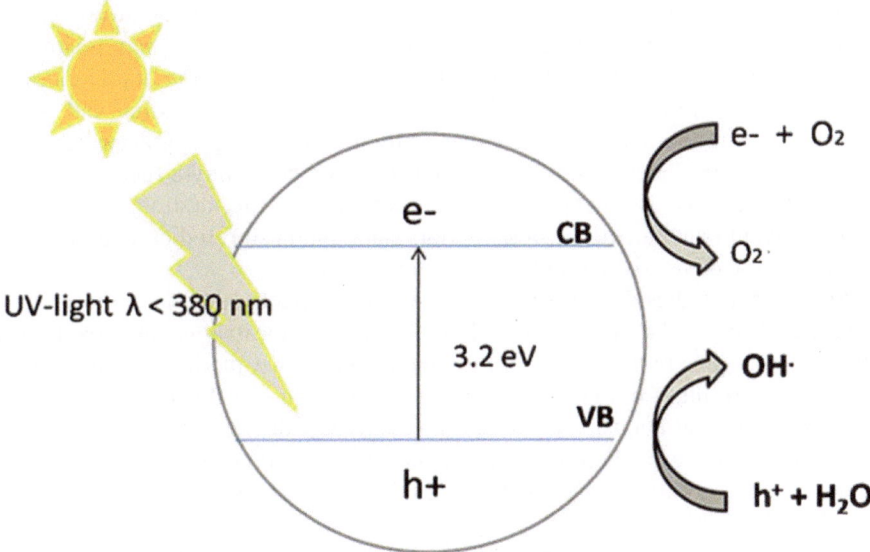

Fig. 7.1 Schematic illustration of TiO_2 electronic structure characterised by its valence (VB) and conduction band (CB) energy positions. Reprinted from Fonseca et al. (2010), copyright 2010, with permission from Elsevier

photocatalytic properties of TiO_2 determines a superhydrophilic interface that allows water to form a thin film on the solid surface, which causes a better wetting of the contaminants (Graziani et al. 2013).

Recently, many studies have been carried out concerning the TiO_2 benefits against building biodeterioration, but the algicidal effect of TiO_2-loaded cementitious materials has been investigated only in very few cases.

For example, concerning cement surfaces, it was observed that TiO_2 treatments lead to a 66 % reduction in the growth of algae in comparison to unprotected cement surface and this effect could be increased to 87 % by adding 1 % of a noble metal such as Pt or Ir (Linkous et al. 2000).

In order to prevent algal fouling on existing structures, Maury-Ramirez et al. (2013) analysed a commercial TiO_2 coating compared to an innovative TiO_2 one which combines water repellency and photocatalytic properties. They evaluated the efficacy by means of an accelerated algal growth test set-up with UV-A irradiation and using *Chlorella vulgaris* as algal species. During 16 weeks, monitoring of these TiO_2-based strategies was conducted by visual inspections, determination of the algal coverage (%) and human perception of the colour changes (DE) produced on the samples. Results proved that samples prepared with the commercially available TiO_2 coating did not show visible algal growth and almost no significant coverage. On the contrary, cement paste samples containing 5 and 10 % TiO_2 did not appear to be efficient to avoid algal growth. Although more research should be conducted to enhance the algicidal activity of the innovative TiO_2-based coating, the results obtained (such as significant reduction in the coverage rate and algal coverage (20 % during this test) are promising.

Further studies were aimed at evaluating the efficacy of TiO_2 in preventing algal fouling on mortars. Fonseca et al. (2010) compared different treatments (anatase and two conventional biocides) for preventing biodeterioration of mortars. The treatments were applied both in situ on walls of a real building and in the laboratory on mortar slabs, as a coating or by mixing during mortar preparation. The mortars were inoculated with cyanobacteria and chlorophyta species. Results showed that anatase is a better agent for preventing biodeterioration than the two tested conventional biocides in both mortars slabs and in situ studies, conferring an excellent protective coating and self-cleaning properties.

Some other studies demonstrated a more limited effectiveness of photocatalytic surfaces. Gladis and Schumann (2011) found that photocatalytic surfaces did not affect phototrophic biofilms on roof tiles under outdoor weathering, also in contrast to precedent results from laboratory studies. The causes for this discrepancy may depend on the low intensity UV exposure (northeast direction): this condition is favourable for the growth of microalgae, but it is not optimum for the activation of TiO_2. In order to further investigate this problem, Graziani et al. (2013) assessed the biocide effect of TiO_2 coatings applied on clay brick specimens contaminated by the green alga *Chlorella mirabilis* and the cyanobacteria species *Chroococcidiopsis fissurarum* under weak UV radiation. Results revealed that the material was not able to fully prevent microalgal biofouling, also because of the "shadow effect" caused by the microalgae themselves on the nanofilm, but it efficaciously prevented the adhesion of the microorganisms on the treated substrates by forming of a superficial water film. This property resulted in a good self-cleaning efficiency of TiO_2.

Authors concluded that, when TiO_2 coatings are applied on façades exposed to weak UV radiation (i.e. north façades), they cannot stop biofouling due to the multiplication of microorganisms. Nevertheless, during a washing process, for instance due to wind-driven rain, TiO_2 proved to greatly enhance the self-cleaning ability of surfaces. Further researches are currently in progress to confirm this hypothesis.

Recently, some other specific types of nanoparticles, such as nanosilver, nanocopper, nanozinc oxide, have been developed as additives for the protection of paints, against microbial degradation and physical and chemical deterioration. They protect coatings and are generally not released into the environment, thus avoiding the pollution problems of biocides (Gaylarde et al. 2011).

The most known are silver (Ag) nanoparticles (AgNPs) because of their high antimicrobial properties. They can be synthesised by several physical, chemical, and biological methods, and act in a multidirectional way, giving an advantage over other biocides. Silver inhibits the replication of DNA, disturbs the electrical potential and functioning of the cytoplasmic membranes, which leads to the outflow of many metabolites from the cells, causes loss of bioactivity of amino acids and disrupts respiratory processes (Gutarowska et al. 2012).

Silver is not always added as nanosilver or silver salts to the paints. In many silver-containing paints, the silver has been added as silver zeolite. Zeolites are ceramic solid materials with a three-dimensional grid-like structure resulting in a network of orthogonal pores running throughout each crystal. Paints containing silver zeolite show good antibacterial effects (Kaiser et al. 2013).

These properties of AgNPs have been exploited widely for their use in medicine, textiles, surfaces of polymers, steel, cosmetics, food packaging, filtration of water, renewable energies, environmental remediation and electronic devices, but much wider testing is required before they can be shown to be active against the formation of mixed biofilms on building surfaces.

For some time, the paint industry has been thinking of using nanocopper as biocidal additive in their paints in two forms: the nanoparticulate forms (metals) and the soluble forms (ions). Nanocopper in concentrations as low as 40–60 µg/ml causes toxicity to a variety of organisms such as algae, bacteria, crustaceans and fish (Kaiser et al. 2013).

Zinc oxide nanoparticles are among the most commonly utilised nanomaterials with wide ranging applications. The use of zinc oxide in the paint industry for the protection of façades from microbial growth is quite well established (Gladis et al. 2010). The toxicity of nanozinc oxide is a result of the dissolution and subsequent release of zinc ions that are highly toxic to water organisms (e.g. algae, crustaceans, fish, nematodes, ciliates and bacteria), as already seen with other metallic nanomaterials. The soluble form of zinc oxide (ionic zinc) is much more toxic than its metallic form for both nanosilver and nanocopper.

Zhang et al. (2013) incorporated two highly effective photocatalytic nano-sized metal oxides (titanium dioxide and zinc oxide) into silane/siloxane based water repellents to further enhance the biofouling resistance of treated mortar slabs. Results from a controlled culture streaming study (with blue-green and green algae) showed that biofouling intensity and surface coverage was significantly reduced by using nanoparticulate additives when compared to the control treatment and was attributed to the photo-induced breakdown of contaminants. Surface roughness, pore morphology and visual aesthetics were not effectively altered by the treatments.

Nanosized silica dioxide particles are added to paints to improve scratch resistance, to protect the paint against corrosion and to provide the product with a high gloss. They are widely used in wood preservation (hydrophobation) and in coating applications (Kaiser et al. 2013).

Silica nanoparticles in an organic polymer dispersion have been stated to resist to "mould formation" because the hydrophilic surface allows rapid spreading of raindrops, producing a washing effect, as well as rapid drying (Pagliaro et al. 2009). Therefore, this material is widely used in maritime construction and ships, but very few controlled trials of the potential for biodeterioration of these new coatings as paints on terrestrial substrates have been found in the scientific literature.

Today, the use of engineered nanomaterials in paints is still at an early stage of research. The expected benefits have not been fully tested so far. The potential risk is not clear. Early results suggest that the use of engineered nanomaterials in paints will not result in an increase in the level of exposure of nanosized materials and will not result in additional health risk to users more than the conventional production in this field as long as they follow the established practices of prevention and protection.

According to Kahru et al., nanosilver and nanozinc oxide are classified as "extremely toxic", followed by nanocopper, which is classified as "very toxic" (Kahru and Dubourguier 2010). Moreover, the release of toxic nanometals from paints is directly related to the amount of painted surfaces in a certain area. Furthermore, the ability of nanoparticles and ions to adhere to inorganic and organic particles to form larger aggregates favours their transfer from the water column into the sediments, which leads to a significant decrease in their bioavailability and toxicity (Kaiser et al. 2013).

At the same time, there is no guarantee that nanomaterials will achieve the proposed benefits, such as water repellence, scratch resistance, improved durability and prevention of biological growth on a long-term perspective since there are no long-term studies available. Nanosilver in paints is easily washed out by rain. Photocatalytic-active nanotitanium dioxide can accelerate the photocatalytic degradation of the paint matrix.

Another limitation to rapid marketing of such products is constituted by their cost: silver is a precious metal and therefore very costly, nanocopper has similar antimicrobial properties as nanosilver and is much cheaper but still quite expensive.

Titanium dioxide is cheap compared to other nanomaterials and therefore has been used as a whitener pigment in paints for decades. Moreover, its photocatalysis property clashes with the fact that on the facades of buildings the dust settles so that it will be reduced after a certain time.

In the future, nanomaterials could replace biodegradable biocides and improve the paint properties as well as impede colonisation by microorganisms. The safety, effectiveness and sustainability of these products are the key elements to be investigated for further research for their marketing and distribution.

7.3 Prevention and Control Strategies in NZEB

The first priority of an NZEB is energy efficiency, in the phase of construction and use. Since the main factors that lead to the development of microorganisms in buildings can be attributed to environmental conditions, design features and material properties of substrates, prevention and control strategies in NZEB should be established in order to mitigate or even eliminate the causes and conditions of their growth by using energy efficient and environmental friendly methods.

Concerning the proliferation of mould on the inner surfaces of the buildings, it is mainly a result of a critical level of RH and internal temperature and a specific capacity of the material to retain water (adsorption curve). The phenomenon is exacerbated by lack of air exchange between internal and external environment, a characteristic of modern buildings with air-sealed and highly insulated envelope (see 6.3). Therefore, it is clear that to prevent the development of moulds, it is necessary to act on two different fronts:

- control of indoor climatic conditions;
- choice of specific finishes and coating materials.

For the control of indoor climatic condition, not only the way of life of the inhabitants of a building is important but equally important is the plant and equipment that is in it.

Concerning the occupants' behaviour, some good practices and advice come from Khan and Mohan Karuppayil and mainly concern the way to use and maintain plant equipment (for instance, frequent cleaning of the A/C filters) (Haleem Khan and Mohan Karuppayil 2012). Further steps to reduce moisture include keeping potted plants clean, increasing bathroom and kitchen ventilation, to keep houseplants healthy, keeping the moisture sensitive materials dry, and keeping carpets clean. Wallpaper or panelling may be removed to reduce the severity of fungal contamination.

Theoretically, residents of homes could avoid condensation by keeping rooms at a constant, steady, low temperature, but this is impossible in practice. In a moderate climate, hygiene ventilation is responsible for about half or more of the energy costs in well-insulated dwellings. Consequently, this field represents a massive gross energy saving potential (Laverge et al. 2011).

Since the problem is often moisture and not temperature, the more direct solution is to use technology designed to remove moisture rather than to keep the temperature steady.

Mechanical ventilation systems in houses, now generally required by national standards (ANSI/ASHRAE 2003) as by other requirements, are increasingly viewed as a technology that will complement tighter building envelopes and address the need for additional indoor moisture control. While mechanical ventilation systems can be potentially effective in providing controlled air exchange and in helping to control moisture, these systems require proper selection, design, installation and operation to be effective.

Dehumidifiers are a widely known device and can be readily purchased. Using a dehumidifier has proved to be a way to solve the problem of condensation, presumably leading to far less mould formation. Galvin demonstrated that using a dehumidifier offers considerable energy saving compared to using a home heating system to achieve the same goal, i.e. of reducing condensation and mould formation and growth (Galvin 2010). He also suggests developing a 'smart' dehumidifier system, with temperature sensors on the indoor window surfaces. When the temperature here falls below a specific minimum temperature, the target humidity would automatically reset to a lower level and the dehumidifier will switch on. This could reduce running costs while achieving optimal moisture reduction.

A proven solution, above all in cold climates, is to install a heat-exchange ventilator system, in which 'fresh' incoming air is heated by 'stale' outgoing air in a capillary system. This provides a constant interchange of air between indoors and outdoors without wasting heat.

However, in moderate climates, the payback time for investments in heat recovery ventilation are long, especially in buildings with relatively low air change rates such as dwellings. Due to its competitive price setting as well as due to reports in popular media and scientific literature about possible health risks associated with heat recovery systems, simple mechanical exhaust ventilation dominates the residential ventilation market in these climatic regions (Laverge et al. 2011).

Furthermore, the introduction of HVAC systems and sophisticated equipments providing an adequate mechanical ventilation seems not to be the best solution for small, low-density interiors, where they would be a source of noise and where the inhabitants would be obliged to keep the windows closed with not negligible aesthetic and psychological consequences.

Even if with HVAC devices it is possible to provide an acceptable indoor climate, nevertheless, there is a desire to develop more passive and less energy intensive methods of moderating the indoor environment. These passive methods are gaining popularity because they are energy conscience and environmentally friendly.

Hygrothermal assembly design, by studying structures, materials and critical joints, is an essential element for preventing moisture damage and guaranteeing longer service life for buildings.

A promising strategy which acts directly on the internal finishing, is related to the use of "moisture buffering" materials to dampen indoor humidity variations.

The interior humidities are caused by interior moisture sources, moisture transport by ventilation air and moisture exchange with the room enclosures and interior objects. The "moisture buffering" effect is known as the capacity of the interior finishing and furnishing materials to moderate indoor humidity in buildings, thanks to their hygroscopic ability. Moisture buffering materials are able to adsorb and desorb moisture from the adjacent air and can be used to control indoor humidity variations without additional energy costs (Janssen and Roels 2009).

Several authors stress the importance of moisture buffering in the global interior humidity managing, supported by measurements (Cerolini et al. 2009; Steeman et al. 2010a) and simulations (Janssen and Roels 2009; Steeman et al. 2010b).

Researchers have shown that several materials used in building construction— cellular concrete, bricks, wood and wood-based materials (Hameury 2005; Hameury and Lundström 2004; Kuenzel et al. 2004; Osanyintola et al. 2006) and cellulose insulation (Padfield 1999)—or infurniture and furnishings (textiles, wood and paper) (Svennberg et al. 2004) show a moisture buffering behaviour.

Interior moisture buffering is shown to positively affect energy consumption, component durability, thermal comfort and air quality.

Also concerning microorganism development on the external finishing of walls, "passive" strategies acting on the properties of materials themselves could be employed.

As already observed (6.2), walls with external insulation systems are widely affected by microbial growth because the low thermal mass of the exterior render combined with high thermal resistance of the insulation layer leads to frequent overcooling and consequent condensation of the external surface by long-wave radiation exchange with the sky during the night.

A good way to prevent microorganism growth on these elements is then to reduce the frequency of condensation by limiting the periods of overcooling. This could be achieved by different methods:

- increasing the thermal inertia of the exterior render by adding phase change materials (PCM);
- reducing the night-time irradiation to the sky by applying infrared reflecting paint coats (Sedlbauer et al. 2011);
- reducing the surface absorption capacity of the substrate and avoiding materials which are characterized by surface roughness since they are able to increase moisture retention.

The use of infrared reflecting coatings and PCM into the external layers are strategies which have already been examined in several investigations (Krus et al. 2006; Kuenzel and Sedlbauer 2007; Künzel 2010); the potential of both options to solve the exterior condensation problem looks promising.

The insertion of phase-change materials (PCM) can be achieved into the ETICS external layer, usually composed of a reinforced stucco base coat and a finish coat, or directly into the external layer of the insulation slabs. PCM can enhance the thermal storage capacity of the facade, in order to store the heat from the daily solar warming of the building component. The heat of fusion will slow down the cooling process, thus keeping surface temperature above the dew point. Paraffin for example, which is available with various melting point ranges, can serve as PCM.

Another possibility to increase the heat capacity of an ETICS render is to use a thicker plaster. Normally, a system with 3 mm plaster is applied. Nevertheless, researches carried out on a test façade with a thicker plaster (10 mm) have demonstrated that the effect of a thicker plaster was not much more than a "time shift" on surface temperature decreasing of circa 1 h in comparison to standard construction (Lengsfeld and Krus 2004).

To speak of infrared-reflecting films or coatings, the reflected ratio of infrared radiation must be clearly higher than 10 %. In the case of opaque layers of building components, the ratios of emitting and reflecting radiation of a certain wavelength always complement one another to 100 %: the higher the long-wave reflection, the lower the emission. This is why those layers are often called "low-e", meaning low IR emittance (Kuenzel and Sedlbauer 2007). The reduction in the long-wave infrared (IR) exchange of radiation of building surfaces with their environment by means of infrared-reflecting layers can contribute to reduce the temperature drop below the ambient air temperature in the night-time and the consequent dew formation.

Reflective paints are also recognised as a fundamental strategy that dense urban areas can deploy at low-cost to mitigate the "heat island" effect, which is the environmental overheating that occurs in urban areas due to the absorption of solar radiation by the surfaces of buildings, roads, etc. (Akbari et al. 2008, 2001; Synnefa et al. 2007).

The most important technology in the production of cool paints is the formulation with complex inorganic coloured pigments (CICPs) or mixed metal oxide (MMO) pigments (Uemoto et al. 2010). Paints containing these pigments have been shown to reduce the duration of dawn condensation on facades, decreasing growth of fouling microorganisms by approx. 15 % (Sedlbauer et al. 2011).

Surface temperature on a concrete wall with ETICS ($U = 0.35$ W/m^{2}K) and an IR coating, investigated by Lengsfeld and Krus was clearly warmer during the night than on a façade without an IR coating (Lengsfeld and Krus 2004). The consequent Time of Wetness was lower.

The performance of novel rendering systems including PCM additives and/or Low-e coating has been extensively investigated by Sedlbauer et al. (2011) during field tests as well as by hygrothermal simulations. Simulations results showed that a thicker external plaster with a high thermal mass can reduce dew point temperature undercut by a maximum of 20 %, IR paint by almost 30 % and a latent thermal storage layer even by 70 %. The combination of both measures (PCM + IR) can further reduce the duration of dew point temperature undercut, as an extreme case. Field results confirmed that the combination of PCM plaster and IR paint is able to reduce exterior condensation by almost 50 %, which is clearly a higher reduction than the results with PCM additive (30 %) and IR paint alone (15 %).

Nevertheless, before these scientific solutions could be turned into marketable products, there are still many problems to be solved. One issue is their long-term performance and appearance.

The effect of reflective coating products could only last over a short period of time because the albedo value is altered by the growth of microorganisms, premature staining, alterations in appearance and dust deposition.

Studies by Bretz and Akbari showed that most of the albedo degradation of coatings occurred within the first year of application, and even within the first 2 months of exposure (Bretz and Akbari 1997).

Besides, the fact that IR-reflecting layers are frequently very vapour-tight due to their metal content can be a problem in terms of moisture protection. Yet, there are recent developments in the field of breather membranes combining IR reflection and high water-vapour permeability (Kuenzel and Sedlbauer 2007).

Stucco with PCM is usually obtained by adding microencapsulated hydrocarbons, which are very hydrophobic and which diminish the coherence of the stucco during application leading to durability and workability problems. Another important issue of PCM is the temperature range where the phase change phenomenon takes place (melting point): it must be set just above the dew point of the ambient air to be effective. Since outdoor-air dew point depends on climate and season, a lot of optimisation work is required (Künzel 2010). The optimisation of the phase change point can be calculated by applying a reference year. Yet, the differences in climate conditions in subsequent years can be so immense that not even this kind of optimised PCM can be continuously effective.

Finally, an effective and simple good practice for limiting biodeterioration is the maintenance of building elements: the periodic simple cleaning of exposed surfaces eliminate the "soiling" effect due to the deposit of particles, which can favour the development of reproductive bodies.

References

Akbari H, Pomerantz M, Taha H (2001) Cool surfaces and shade trees to reduce energy use and improve air quality in urban areas. Sol Energy 70:295–310. doi:10.1016/S0038-092X(00)00089-X

Akbari H, Menon S, Rosenfeld A (2008) Global cooling: increasing world-wide urban albedos to offset CO_2. Clim Change 94:275–286. doi:10.1007/s10584-008-9515-9

ANSI/ASHRAE (2003) STANDARD 62.2-2003 Ventilation and acceptable indoor air quality in low-rise residential buildings

Balzarotti-Kammlein R (1999) An innovative water-compatible formulation of Algophase® for treatment of mortars. In: Proceedings of the international conference on microbiology and conservation of microbes and art, Florence, pp 16–19

Bester K, Lamani X (2010) Determination of biocides as well as some biocide metabolites from facade run-off waters by solid phase extraction and high performance liquid chromatographic separation and tandem mass spectrometry detection. J Chromatogr A 1217:5204–5214. doi:10.1016/j.chroma.2010.06.020

Bretz SE, Akbari H (1997) Long-term performance of high-albedo roof coatings. Energy Build 25:159–167. doi:10.1016/S0378-7788(96)01005-5

Cerolini S, D'Orazio M, Di Perna C, Stazi A (2009) Moisture buffering capacity of highly absorbing materials. Energy Build 41:164–168. doi:10.1016/j.enbuild.2008.08.006

European Parliament (1998) Directive 98/8/CE of the European Parliament and of the Council of 16 February 1998 concerning the placing of biocidal products on the market

European Parliament (2004) Directive 2004/42/CE of the European Parliament and of the Council of 21 April 2004 on the limitation of emissions of volatile organic compounds due to the use of organic solvents in certain paints and varnishes and vehicle refinishing products and amendi

European Parliament (2008) Regulation (EC) No 1272/2008 of the European Parliament and of the Council of 16 December 2008 on classification, labelling and packaging of substances and mixtures, amending and repealing Directives 67/548/EEC and 1999/45/EC, and amending Regulation (EC)

Fonseca AJ, Pina F, Macedo MF et al (2010) Anatase as an alternative application for preventing biodeterioration of mortars: evaluation and comparison with other biocides. Int Biodeterior Biodegradation 64:388–396. doi:10.1016/j.ibiod.2010.04.006

Galvin R (2010) Solving mould and condensation problems: a dehumidifier trial in a suburban house in Britain. Energy Build 42:2118–2123. doi:10.1016/j.enbuild.2010.07.001

Gaylarde CC, Morton LHG, Loh K, Shirakawa MA (2011) Biodeterioration of external architectural paint films—a review. Int Biodeterior Biodegradation 65:1189–1198. doi:10.1016/j.ibiod.2011.09.005

Gladis F, Schumann R (2011) Influence of material properties and photocatalysis on phototrophic growth in multi-year roof weathering. Int Biodeterior Biodegradation 65:36–44. doi:10.1016/j.ibiod.2010.05.014

Gladis F, Eggert A, Karsten U, Schumann R (2010) Prevention of biofilm growth on man-made surfaces: evaluation of antialgal activity of two biocides and photocatalytic nanoparticles. Biofouling 26:89–101

Graziani L, Quagliarini E, Osimani A et al (2013) Evaluation of inhibitory effect of TiO_2 nanocoatings against microalgal growth on clay brick façades under weak UV exposure conditions. Build Environ 64:38–45. doi:10.1016/j.buildenv.2013.03.003

Gutarowska B, Skora J, Zduniak K, Rembisz D (2012) Analysis of the sensitivity of microorganisms contaminating museums and archives to silver nanoparticles. Int Biodeterior Biodegradation 68:7–17. doi:10.1016/j.ibiod.2011.12.002

Haleem Khan AA, Mohan Karuppayil S (2012) Fungal pollution of indoor environments and its management. Saudi J Biol Sci 19:405–426. doi:10.1016/j.sjbs.2012.06.002

Hameury S (2005) Moisture buffering capacity of heavy timber structures directly exposed to an indoor climate: a numerical study. Build Environ 40:1400–1412. doi:10.1016/j.buildenv.2004.10.017

Hameury S, Lundström T (2004) Contribution of indoor exposed massive wood to a good indoor climate: in situ measurement campaign. Energy Build 36:281–292. doi:10.1016/j.enbuild.2003.12.003

ISO/TS 27687 (2008) Nanotechnologies—Terminology and definitions for nano-objects—Nanoparticle, nanofibre and nanoplate

Janssen H, Roels S (2009) Qualitative and quantitative assessment of interior moisture buffering by enclosures. Energy Build 41:382–394. doi:10.1016/j.enbuild.2008.11.007

Kahru A, Dubourguier H-C (2010) From ecotoxicology to nanoecotoxicology. Toxicology 269:105–119. doi:10.1016/j.tox.2009.08.016

Kaiser J-P, Zuin S, Wick P (2013) Is nanotechnology revolutionizing the paint and lacquer industry? A critical opinion. Sci Total Environ 442:282–289. doi:10.1016/j.scitotenv.2012.10.009

Krus M, Fitz C, Holm A, Sedlbauer K (2006) Prevention of algae and mould growth on facades by coatings with lowered long-wave emission. Report, Fraunhofer Institut Bauphysik, Stuttgart, Germany

Kuenzel H, Sedlbauer K (2007) Hygrothermal effects of infrared-reflecting layers. Report, Fraunhofer Institut Bauphysik, Stuttgart, Germany

Kuenzel H, Holm A, Sedlbauer K et al (2004) Moisture buffering effects of interior linings made from wood or wood based products. Investigations commissioned by Wood Focus Oy and the German Federal Ministry of Economics and Labour

Künzel HM (2010) Factors determining surface moisture on external walls. Thermal performance of the exterior envelopes of whole buildings XI international conference, Clearwater Beach, FL

Laverge J, Van Den Bossche N, Heijmans N, Janssens A (2011) Energy saving potential and repercussions on indoor air quality of demand controlled residential ventilation strategies. Build Environ 46:1497–1503. doi:10.1016/j.buildenv.2011.01.023

Lengsfeld K, Krus M (2004) Microorganism on façades-reasons, consequences and measures. Report, Fraunhofer Institute for Building Physics, Holzkirchen, Germany

Linkous CA, Carter GJ, Locuson DB et al (2000) Photocatalytic inhibition of algae growth using TiO 2, WO 3, and cocatalyst modifications. Environ Sci Technol 34:4754–4758. doi:10.1021/es001080+

Maury-Ramirez A, De Muynck W, Stevens R et al (2013) Titanium dioxide based strategies to prevent algal fouling on cementitious materials. Cement Concr Compos 36:93–100. doi:10.1016/j.cemconcomp.2012.08.030

Osanyintola OF, Talukdar P, Simonson CJ (2006) Effect of initial conditions, boundary conditions and thickness on the moisture buffering capacity of spruce plywood. Energy Build 38:1283–1292. doi:10.1016/j.enbuild.2006.03.024

Padfield T (1999) Humidity buffering of interior spaces by porous, absorbent insulation. Report, Department of Structural Engineering and Materials, Technical University of Denmark, Lyngby, Denmark

Pagliaro M, Ciriminna R, Palmisano G (2009) Silica-based hybrid coatings. J Mater Chem 19:3116. doi:10.1039/b819615j

Sedlbauer K, Krus M, Fitz C, Künzel H (2011) Reducing the risk of microbial growth on insulated walls by PCM enhanced renders and IR reflecting paints. XII DBMC international conference on durability of building materials and components, Porto, Portugal

Shirakawa MA, Gaylarde CC, Gaylarde PM et al (2002) Fungal colonization and succession on newly painted buildings and the effect of biocide. FEMS Microbiol Ecol 39:165–173. doi:10.1111/j.1574-6941.2002.tb00918.x

Shirakawa MA, Tavares RG, Gaylarde CC et al (2010) Climate as the most important factor determining anti-fungal biocide performance in paint films. Sci Total Environ 408:5878–5886. doi:10.1016/j.scitotenv.2010.07.084

Som C, Wick P, Krug H, Nowack B (2011) Environmental and health effects of nanomaterials in nanotextiles and façade coatings. Environ Int 37:1131–1142. doi:10.1016/j.envint.2011.02.013

Steeman M, Van Belleghem M, De Paepe M, Janssens A (2010a) Experimental validation and sensitivity analysis of a coupled BES–HAM model. Build Environ 45:2202–2217. doi:10.1016/j.buildenv.2010.04.003

Steeman M, Janssens A, Belleghem MV, De Paepe M (2010b) Validation of a coupled BES-HAM model with experimental data. In: Proceedings of the 1st Central European symposium on building physics, Cracow, Poland

Svennberg K, Hedegaard L, Rode C (2004) Moisture buffer performance of a fully furnished room. In: Proceedings (CD) of the performance of exterior envelopes of whole buildings IX international conference, Clearwater Beach, FL

Synnefa A, Santamouris M, Akbari H (2007) Estimating the effect of using cool coatings on energy loads and thermal comfort in residential buildings in various climatic conditions. Energy Build 39:1167–1174. doi:10.1016/j.enbuild.2007.01.004

Tiano P (2002) Biodegradation of cultural heritage: decay mechanisms and control methods. 9th ARIADNE workshop "Historic Material and their Diagnostic", ARCCHIP, Prague, Czech Republic

Uemoto KL, Sato NMN, John VM (2010) Estimating thermal performance of cool colored paints. Energy Build 42:17–22. doi:10.1016/j.enbuild.2009.07.026

Urzì C, De Leo F (2007) Evaluation of the efficiency of water-repellent and biocide compounds against microbial colonization of mortars. Int Biodeterior Biodegradation 60:25–34. doi:10.1016/j.ibiod.2006.11.003

Urzì C, Leo F De, Galletta M et al (2000) Efficiency of biocide in "in situ" and "in vitro" treatment. Study case of the "Templete de Mudejar", Guadalupe, Spain. In: Proceedings of the ninth international congress on deterioration and conservation of stone, Venice, Italy, pp 531–539

Wangler TP, Zuleeg S, Vonbank R et al (2012) Laboratory scale studies of biocide leaching from façade coatings. Build Environ 54:168–173. doi:10.1016/j.buildenv.2012.02.021

Warkentin M, Schumann R, Messal C (2007) Faster evaluation. Eur Coat J 09:26

Zhang Z, MacMullen J, Dhakal HN et al (2013) Biofouling resistance of titanium dioxide and zinc oxide nanoparticulate silane/siloxane exterior facade treatments. Build Environ 59:47–55. doi:10.1016/j.buildenv.2012.08.006

Chapter 8
Conclusions

The problem of the proliferation of microorganisms in buildings has a profound effect on the lifestyle of the people, with health, economic and social consequences.

These organisms populate for a very long time on the buildings envelope and indoors. However, their presence has recently been exacerbated by contemporary way of constructing buildings that are directed toward a Nearly Zero Energy standard. In fact, the need to reduce energy consumption and greenhouse gas emissions has actually led many countries to adopt legislation and strategies to improve the thermal performance of buildings, mainly by reducing the thermal transmittance and air permeability of the envelope. This "sealing action" aims to minimize air leaks and consequent heat dispersions as much as possible.

The air tightness of the internal environment, if not properly managed with ventilation equipment, may lead to high internal moisture load and consequent surface condensation. A very thick insulation material on the exterior side of the envelope is more subject to exterior undercooling phenomena and consequent condensation. Moisture loads and surface condensations are the most favourable conditions for the proliferation of microorganisms, such as algae and fungi.

Consequently, despite improvement in building energy efficiency and better quality requirements for living spaces, over the last few decades the number of reports on the presence of microorganisms on building facades and indoors is still increasing.

As a result, the aesthetic quality and durability of buildings could be seriously impaired. Algae and mould on facades contribute to the defacement of paint and finishes. Mould growth indoors could become responsible for several types of illnesses and pathologies experienced by building occupants. These illnesses and pathologies are grouped under the name of "Sick Building Syndrome".

In order to preserve buildings from the colonization of microorganisms, it is of primary importance that we have a better understanding of the principal proliferating organisms and of their main growth conditions. Moreover, there is a growing demand for calculation methods in building engineering to assess the moisture behaviour of building components and microorganism risk prediction in order to ensure a healthy environment, and to avoid material defacement, with social and economic consequences.

E. Di Giuseppe, *Nearly Zero Energy Buildings and Proliferation of Microorganisms*, 93
SpringerBriefs in Applied Sciences and Technology, DOI: 10.1007/978-3-319-02356-4_8,
© The Author(s) 2013

Many building hygrothermal analysis methods are able to simulate the coupled transport processes of heat and moisture for one or multidimensional cases which aim to predict biological risks. Nevertheless, a more in-depth study of the growth of microorganisms under transient conditions is still required in order to be able to define the most reliable prediction model. Other further improvements could be to develop prediction models that include a spread in germination time and growth rate, a variation in material properties, moisture load, transfer coefficients and bad workmanship. To do so, additional measures in laboratory and on real buildings and components would be desirable.

Controlling and preventing the proliferation of microorganisms in buildings have become more pressing and important. Among traditional solutions, chemical treatments do not ensure long-term protection since they need re-application over time and are likely to be effective at the expense of durability and aesthetics of the substrate material. Among them, water repellents and biocides are commonly used, but nowadays it is important to control biodeterioration process with new environmentally friendly technologies.

The advantages of nanomaterials are that they can be fixed more effectively in the coating matrix, reducing their leaching out into the surrounding environment. Nevertheless, they are still at an early stage of research. The expected benefits have not been fully tested yet. Their potential health effects and ecological risk are also still not clear.

Other researches are looking towards safer and more sustainable strategies linked to a proper design, choice of materials and construction of buildings, aiming to obtain a good envelope performance without any added costs. Concerning interior finishing materials, "moisture buffering" is a very promising method which acts directly on substrates to dampen indoor humidity variations.

As regards external finishing materials, the potential of using infrared reflecting coatings and PCM to solve the exterior condensation problem, by reducing the night-time irradiation to the sky and enhancing the thermal storage capacity of the façade, looks encouraging.

In the future, all these strategies should be further investigated in order to replace chemical biocides and to improve the properties of materials. Their safety, effectiveness and sustainability are the key elements to be investigated for their real application, marketing and distribution.

Acknowledgments The multidisciplinary topic of this book has necessitated the cooperation of experts in different disciplines. Chapter 6 has been written in collaboration with Prof. Marco D'Orazio, whose expertise in building physics was substantial for the definition of thermo-physical phenomena that regulate the heat and moisture transport in NZEB and their relationship with biological growth. Prof. Enrico Quagliarini gave a significant contribution in Chap. 7, for the description of remedial actions and future innovations. The author is very grateful to them. The author wishes also to acknowledge Prof. Francesca Clementi, Dr. Lucia Aquilanti and Dr. Andrea Osimani for their assistance, support and scientific guidance in writing Chaps. 3 and 4.